PRAISE FOR *T*

'... a moving and sadly accura[
greatest rive[
The Age

'Timely and revealing ... an immensely readable travelogue, revelling
in the rich heritage and character of the Australian bush.'
Australian Bookseller & Publisher

'[Hammer's] clear-eyed analysis of the environmental issues is
matched by compelling and often poignant storytelling.'
G Magazine

'This is an important book ... Hammer has brought us the humanity
of the people and families so badly affected.'
Good Reading

'A disturbing and though-provoking account for all Australians.'
Sun Herald

'[Hammer's new work] ... deserves a place among the
Australian classics'
Tony Wright, *The Age*

'Poignant, urgent environmental writing at its best.'
Better Homes and Gardens

'*The River* should be required reading for all politicians ... because it
is a true and honest account of a great Australian icon in crisis.'
The Canberra Times

THE
RIVER

CHRIS HAMMER

THE RIVER

A Journey through the Murray-Darling Basin

MELBOURNE
UNIVERSITY
PRESS

Melbourne University Publishing acknowledges the traditional owners of the unceded land on which we work, learn and live: the Wurundjeri Woi-wurrung peoples of the Kulin nation. We pay respect to elders past, present and future, and acknowledge the importance of Indigenous knowledge.

MELBOURNE UNIVERSITY PRESS
An imprint of Melbourne University Publishing Limited
Level 1, 715 Swanston Street, Carlton, Victoria 3053, Australia
mup-info@unimelb.edu.au
www.mup.com.au

First published 2010
Second edition published 2011
This edition published 2025
Text © Chris Hammer, 2010, 2025
Photographs © Chris Hammer, 2010
Design and typography © Melbourne University Publishing Limited, 2010

Text design by Phil Campbell
Cover design by Nada Backovic
Typeset by J&M Typesetting
Printed in Australia by McPherson's Printing Group

A Cataloguing-in-Publication entry is available from the National Library of Australia.

A catalogue record for this
book is available from the
National Library of Australia

9780522881424 (paperback)
9780522888162 (ebook)

For Tomoko, Cameron and Elena

Contents

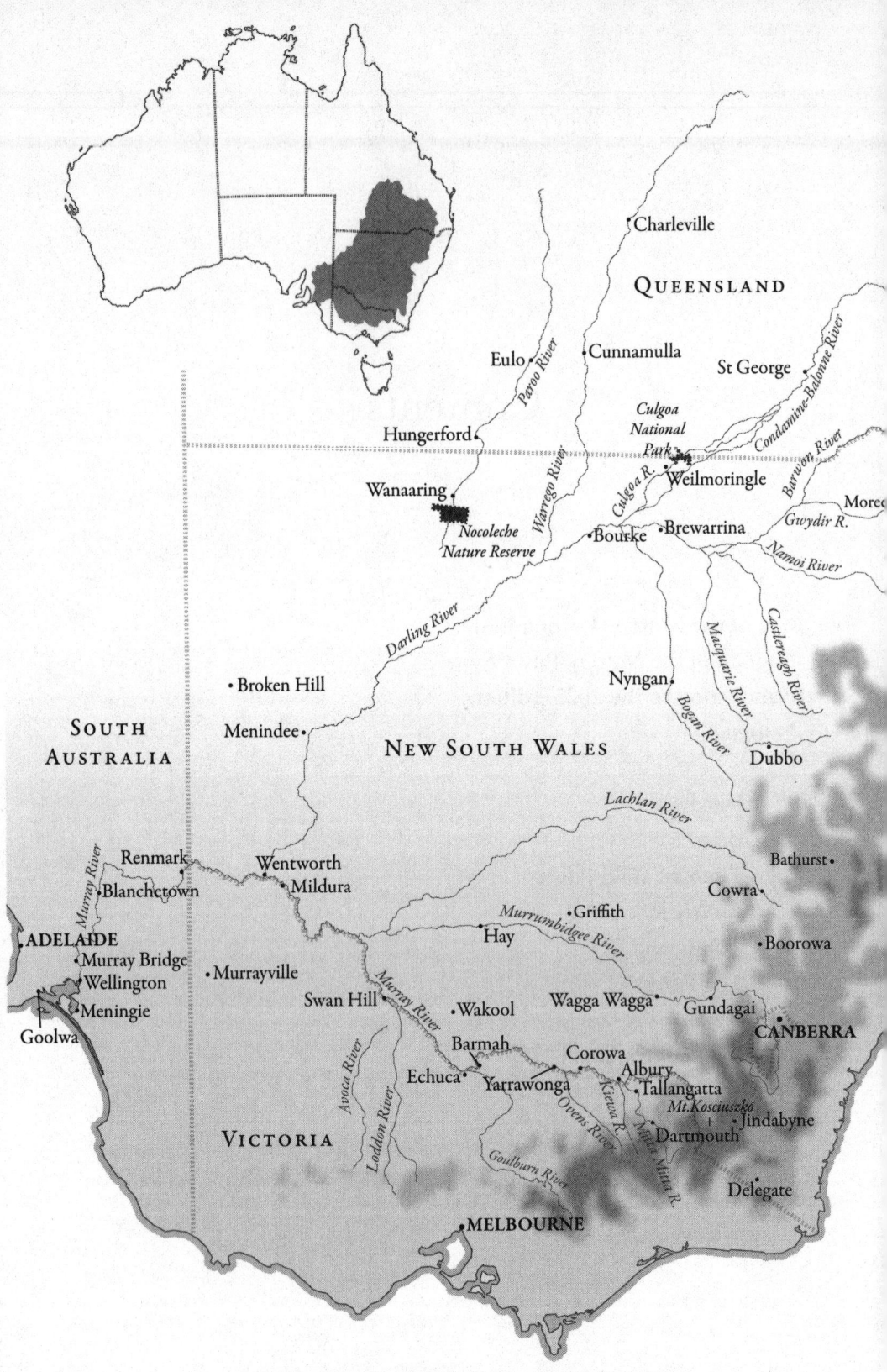

The towns and rivers of the Murray-Darling Basin.
(Guy Holt Illustration and Design.)

The Inland Sea: an artificial River Murray comprehensively regulated by dams, weirs and barrages.
(K Guilfoyl, in RMC: The Work of the River Murray Commission, *comp. AF Ronalds, Government Printer,*
Melbourne, 1946. © MDBA.)

Introduction to the 2025 Edition

It's almost seventeen years since I fired up my aged station wagon and ventured into the drought-ravaged Australian interior, the beginning of a journey of discovery, comprehension and bewilderment. I set out to learn about the rivers of the Murray-Darling Basin, the environments they nurture and the people who dwell by their banks. I ended up appreciating so much more: the Indigenous connection to the rivers, the role the bush shaped in forging our national identity, the threat from climate change and the challenges that lie ahead.

I travelled the Murray-Darling from its headwaters in Queensland to its mouth in South Australia at the height of the Millennium Drought. It was the most severe since European settlement, lasting for more than a decade. My journey – actually a series of journeys – took place over the summer of 2008–09, as the great dry reached its zenith. So I saw it at its worst, when irrigators were on their knees, farmers were walking off their properties and the spectre of suicide stalked the river towns. It was a time when redgum forests were dying, ecosystems were collapsing and wildlife numbers were plummeting. The country, having burnt in 2003, burnt once more: 173 dead in Victoria's Black Saturday conflagration. It was feared that South Australia's lower lakes, near the Murray's mouth, would evaporate past the point of no return, their residual water turning to battery acid, and that Adelaide would run out of drinking water.

During 2008 I'd been reporting on the environment for Melbourne's *The Age*, and a big part of my task was covering the newly elected Rudd government's proposals to address climate change, as well as its plan to save the Murray-Darling. But the debate was hijacked by vested interests: statistics were manipulated and weaponised, interstate rivalries were exploited, politicians ramped up their rhetorical sleights of hand. It became a dialogue of the deaf: highly technical, all about licences and allocations and gigalitres. I grew sick of it: I wanted to get out there and see what was actually happening. So I was immensely grateful to Melbourne University Press and its then publisher, Louise Adler, when she gave me the opportunity to write this book. My aim was to set aside the politicians and the lobbyists and the self-serving reports commissioned from expensive consultants. I wanted to find out for myself.

The drought itself was well-established by the time of my travels: it had begun in 1996 and spread and deepened as the second millennium passed into the third. By 2003, seven years in, the drought was proclaimed as the worst in recorded history. That was a year of bushfires: the year I witnessed Canberra burning – almost five hundred houses lost along the capital's western edge, four dead – the Snowy Mountains burning, the Victorian alps. And still the drought deepened. For much of southern Australia, 2006 was the driest on record. Think about that: the driest on record, on the driest inhabited continent, coming on top of a decade of little or no rain. Crops died, forests burnt, topsoil blew away. It was not until late 2010, six months after this book was first published, that that great Pacific pendulum finally swung back from El Niño to La Niña, and floods replaced drought.

I had seen the country at a time when it seemed the drought might never end.

Much of that is now forgotten. Complacency has returned. The dams are full and the rivers are flowing. Three out of the past four years have been cooler, wetter La Niñas, and Australia has greened once more. The farmers are taking off record crops, the redgum forests have revived, wildlife numbers are recovering, at least for most species. But 'cooler' has become a relative term, given that globally, 2024 was the hottest year on record, and the World Meteorological Organization tells us the last ten years are the hottest ten on record. The scientists were right; the climate change deniers were wrong.

For the moment, Australia's luck is holding. As I write this, meteorologists are predicting yet another La Niña forming, the first such formation during the summer months in seventy-five years. The wetter cycle is undoubtably better for Australia; most farmers and rural communities would prefer rain, even if it carries the risk of flooding, to the drawn out torture of drought. So too environmentalists and firefighters. But look across the Pacific and witness the devastation unleashed on California in January 2025: Los Angeles burning, in the middle of winter, the land tinder-dry after years of La Niña-induced drought, the fires turbocharged by gale-force winds driven by a heating atmosphere. Regard California and despair, for at some point the cycle will turn, the pendulum will swing, and it will be Australia bearing the brunt of drought. Just how far and how hard it will swing is anybody's guess. Climate change is lending momentum to this pendulum; the floods more intense, the droughts more severe.

So perhaps this is a good time to revisit my travels during that summer. For what happened then will almost certainly come again. Maybe not the next drought or the one after, but it will come. Even before climate change, Australia experienced devastating dries, such as the Federation drought of the 1890s. The question must be, how much worse might the next one be, and are we better prepared than twenty years ago?

Some things have changed for the better, and some for the worse. The year I travelled the basin coincided with the infancy of the Rudd government, when political consensus on tackling climate change and addressing the problems of the Murray-Darling Basin seemed difficult but achievable. Malcolm Turnbull had replaced Brendan Nelson as opposition leader; political bipartisanship appeared possible. That was before Tony Abbott came into the leadership and deployed climate denial as a political wrecking ball. Nevertheless, the Murray-Darling Basin Plan was passed into law in 2012, full of compromises and inadequacies, but at least something. Under the plan more water is reserved for the environment, new infrastructure has reduced evaporation and seepage, the use and allocation of water has become more efficient. But the plan is inadequate, a half-measure. The Wentworth Group of Concerned Scientists states: 'the Basin Plan falls well short of returning the volumes of water that science has shown are required for a healthy river, vast sums of taxpayer's money has been wasted, and communities' confidence in government has been shattered'.

Two things haven't changed: the Australian Constitution—power over the rivers belongs to the states, and they guard it jealously. Queensland, New South Wales and Victoria still resist water flowing from their rivers downstream into the main channel of the Murray. This leaves the federal government, of whatever political persuasion, with a weak hand, left to coordinate, to harass and to buy compliance with swags of money. The second unchanged is the reliance on rain. The plan will only be stress-tested when the pendulum swings and the next great drought is upon us.

I learnt a lot writing this book, about Australia and about myself. One thing I discovered was that I like writing books, that there is something deeply satisfying in completing a considered and long-term endeavour compared to the reactive immediacy of daily journalism. Here in these pages, you'll find the inspirations and the settings for many of my crime fiction books. Read the chapter on Wakool, and you will find the landscape from *Scrublands*; in the chapter on the Barmah-Millewa Forest you will find the location of *The Tilt*; in the description of water trading and irrigation you will find the seeds for *The Seven*. Just as I couldn't have written this book without the years of journalism that predated it, neither could I have written my crime fiction without first writing *The River* and its companion book *The Coast*.

So come with me. Let me take you back to the western rivers, as they were in that drought-ravaged summer, to appreciate our collective past and glimpse our collective future.

Chris Hammer
February 2025

Prologue

The Finniss River Recreational Boating Destination

I drive north on the highway for about 10 kilometres before turning off onto a secondary road. After months of dry the rain is sweeping inland off the Southern Ocean in intermittent squalls, and I drive slowly, forced to squint past the wipers. A few more kilometres and I turn off again, onto gravel. After months of travelling the river catchments, I've developed a nose for this sort of thing, understanding the lay of the land: where the roads follow ridges, where they curve around them, and where they dip to meet the waterways. I've learnt where to look for the water, where the subtle tug of gravity has taken it, carving gullies and creeks, rivers and streams as it goes. I look for another road, hoping there will be one, that will take me from the engineers' heights to nature's low, from the road to the water. The rain pauses and just such a road appears on my left, as if on cue. Tonkin Road it's called. Not much more than a track, really. But it's taking me towards the water, or towards where it should be, to where it once was—the water that has lain in this landscape for seventy years, and ebbed and flowed for thousands of years before that; the water that has gone.

Six months ago I wouldn't have recognised the significance of the view off to the left as I ease the car down Tonkin Road: just another dry creek bed running through just another field. I continue down to where the road stops. The drizzle returns, the day ironically grey, yet the evidence is clear for all to see. I can't help smiling at the jetties. The three of them look so ridiculous in their empty field, raised 3 metres above the dry creek bed. A sign says: 'Finniss River Recreational Boating Destination ... This facility is provided for picnics and overnight stops for people with recreational craft'. I stand on one of the jetties and look towards the horizon, to where the water must be, but there's little chance of seeing any of it; the diminishing shoreline of Lake Alexandrina, the larger of South Australia's two lower lakes, must be at least 2 kilometres away. The lakes lie at the end of the Murray River, just before it enters the sea. Or rather, just before it used to enter the sea.

I clamber down from the jetty into the creek bed where the black soil lays cracked and exposed, like a farm dam at the end of summer. Something white lying against the black of the soil catches my eye: the broken shell of a freshwater mussel, as incongruous in this farmer's field as the jetties. Deja vu. As I crouch to examine it, the wind drops and I catch the smell: a hint of sulphur. Perhaps the misting rain is interacting with the exposed acid sulphate soil to form sulphuric acid. That's the great fear: that the residual pools of the lakebed will turn to battery acid; that the lower lakes will become irredeemably toxic; that the Murray will be dead before it reaches the sea, even if the water does some day return.

I had thought myself well prepared before I began this journey along the rivers six months ago. I'd been perched up in the Canberra Press Gallery for the best part of a year writing on the environment: collecting the data, speaking to the experts, listening to the political debate—the view from above. I'd become familiar with the Murray-Darling, Australia's largest and most important river system, despite rarely venturing out into it. I'd reported on Garnaut, on government green and white papers, on multibillion-dollar water buybacks, on new deals between the federal and state governments, on the ever-shrinking water allocations for irrigators, on the slow death of the wetlands. I'd watched ringside as politicians engaged in their slippery wrestle for ascendancy, and had them visit my desk to convince me of their well-rehearsed truths. I'd listened to environmentalists, farmers, industrialists and scientists. Every second month, I'd faithfully left Parliament House to visit the headquarters of the Murray-Darling Basin Commission and hear the ever-grimmer drought updates and the arid predictions of the Bureau of Meteorology. I could talk giga-litres and allocations and diversions with the best of them. I knew my way round the facts: that the Murray-Darling is Australia's food bowl, supplying some 40 per cent of our agricultural output and supporting some two million inhabitants, plus providing drinking water to a million more in Adelaide. I also knew the prognosis: Ross Garnaut's warning that unconstrained climate change would end agriculture in the basin and effectively depopulate it within a century. But every now and then I'd gaze westward through the hermetically sealed windows of Parliament House, looking to the sometimes snow-capped Brindabellas, and I'd wonder what was happening on the far side of the range, whether the rivers flowing off towards South Australia could possibly be half as dry as the statistics clogging the desk in front of me.

Not that the information hasn't been useful. The maps taught me the basic geography of the basin. They declare that the Murray-Darling system covers a huge swathe of Australia: one-seventh of the continent's land mass; more than a million square kilometres, almost three times the size of Germany. They show the basin stretching from southern Queensland to contain almost all of New South Wales west of the coastal mountains, much of the Victorian hinterland and a crucial swathe of riverland in the east of South Australia. Some maps divide the system into twenty-three separate river catchments, feeding one into another until they eventually flow together into the South Australian Murray and from there to the Southern Ocean. Other maps divide the system into the Northern and Southern Basins, suggesting that what happens in one might have little or no impact on the other. These maps reveal the Northern Basin to be more than twice as big as the Southern Basin; it feeds the Darling River, the sole connection between the two basins. The Northern Basin begins in Queensland, extending west from the Great Dividing Range along the Condamine-Balonne-Culgoa catchment and the Barwon River. The Barwon becomes the main stem as more rivers join from the east: the Gwydir and the Namoi of northern New South Wales, and, from further south, the Castlereagh, the Macquarie and the Bogan, the water coming from as far south as Bathurst, directly west across the Blue Mountains from Sydney.

The maps display, without explaining why, the rivers' curious ability to change names. Up near one river's headwaters, it's called the Condamine before becoming the Balonne, and eventually the Culgoa. Where the Culgoa joins the Barwon, both names are discarded and the Darling is born. By the time the Darling reaches Bourke in far western New South Wales, it has gathered in all its eastern tributaries. Only the 'outside' rivers remain, the Warrego and the Paroo, which flow down from the Queensland desert to join the Darling south of Bourke. The Darling joins the Murray at Wentworth, not far east of the South Australian border.

Although the Northern Basin is massive, it's the Southern Basin where most of the water and three-quarters of the irrigation is found. It's also where the crisis of drought is most keenly felt, and where the climate change predictions are grimmest. The maps show the major rivers of the Southern Basin flowing from the mountains of the Great Dividing Range: the Lachlan and the Murrumbidgee in

New South Wales, the border-defining Murray, and the shorter but water-rich rivers of Victoria—the Kiewa, the Ovens, the Goulburn, the Campaspe and the Loddon—pushing north-west to join the Murray. But the maps don't necessarily show the underground tunnels of the Snowy Mountains Scheme, diverting water from coastal rivers through the mountains to feed the Murray and the Murrumbidgee. By the time the Murray has reached Mildura it has collected almost all its tributaries. The last to join is the Darling, but very little of the water that flows onwards into South Australia comes down it; most of the water originates in the Australian Alps, the forested range extending down past Canberra to the New South Wales snowfields, and on into Victoria, towards Melbourne. More than half the water in the Murray-Darling Basin comes down the Murrumbidgee, the Goulburn, the Loddon, the Broken and the Murray, originating in a catchment less than a seventh the size of the entire basin.

I began my travels armed with these statistics and a bag full of maps. But maps speak nothing of history and tradition and struggle; statistics contain no pride, no despair, no elation. Maps don't buy you drinks in the front bar and regale you with tales of yesteryear; statistics don't whisper the heartbreak of lost land and eroded culture. They cannot show the beauty of a river red gum framed by the first rays of a summer dawn, or the determination in the face of a fourth-generation farmer fighting the weather with one hand and the big city banks with the other, or the meaning of the bush to a young nation still finding its way in the world. Travel down from the mountains and out across the western plains and maps become landscape, statistics become people, and rivers become metaphors. It was the rivers that first led us away from the coast and into the heartland. Before the highways and the railways, they were the conduits that stretched out into the unknown, leading the explorers and the settlers off into the bush, where they searched for water and gold and carried back the seeds of a new identity. And for millennia before that, the rivers, carved out of the landscape by the rainbow serpent, had lain at the centre of Aboriginal life—spiritual and temporal. Rivers water our crops but they also feed our sense of self: the bush ethos that helped shape Australia resonates still, and what happens west of the mountains still says something about who we are, even now. The decline in the Murray-Darling is not just about agricultural output and endangered ecosystems, nor is it just about maps and statistics; it's about us.

I've travelled a circuitous, on-again off-again route to reach the Finniss River Recreational Boating Destination. I started up on the border between Queensland and New South Wales, where dams and diversions have sucked some rivers dry while filling others. I went to Bourke, paying my respects to the crumbling legacy of bush legends and the fabrication of a young nation's identity. I travelled out to the Paroo, the last wild river, and found myself captivated. I journeyed to Menindee, where white logic is once again threatening black spirituality, and to the Murray, where I rejoiced in the most remarkable engineering feats, civil and social, and then on to the Barmah Forest, where I felt the residual magic of a red gum wetland. In Wakool I was welcomed by a community staring fate in the eye. On the South Australian Murray I saw a desert turned green, and a river turned slow and salty. And then I came to the lower lakes, where the combined residue of twenty-three catchments pools together, and we can see reflected in the receding waters just what we have wrought. And finally that journey has taken me here.

I climb back up the bank and out onto one of the jetties. Just three years ago the water here would have been between 1 and 2 metres deep, enough to moor a runabout. The signs and the jetties are all new, all well maintained. There's nothing old or decayed. There's just no water. I knew this would be here, this or something like it, but to see it, to experience it, makes it real. After half a year of intermittent travel, it stills shocks me. Australia's major river system is collapsing. Parts of it are dying; parts of it are already dead. Places like this, near the shores of South Australia's lower lakes, may never recover. Down here, the Murray has stopped flowing; Australia's most significant river no longer reaches the sea. Ecosystems are dying, farmers are going to the wall, towns are emptying. I stand for a few moments longer on the jetty and look out into the dim autumn light and wonder once again how it has come to this.

HEADWATERS

Alan Thompson's bridge spans the Culgoa River at Caringle

It's spring in Canberra, and before I head off to explore Australia's rivers I give the lawn its annual mow. A smattering of seasonal rain has greened up what remains of its once comprehensive cover, and neighbourly guilt compels me to wake the ancient Rover from hibernation. After pulling it from the garage, I pour a litre or two of petrol down its parched gullet and give the pull-start a couple of hefty yanks, praying the engine will catch and I won't be condemned to cleaning the spark plug or mucking out the carburettor. The gods are with me and the elderly machine coughs into life with spluttering protestations and a halo of blue smoke.

This is one of the hidden benefits of drought for those of us who dislike cutting lawns and the cult of the English garden. Can there be a more comprehensive waste of time or money? All those scarce raw materials; all those carbon emissions; all that money and labour devoted to a machine whose sole purpose is to impose uniformity upon suburbia. I'm the family black sheep in this regard. By contrast, my father has always been fond of a therapeutic mow, finding in that cocoon of noise, dust and sweat something at times meditative, at others cathartic. He must find some enjoyment in it; he does it often enough. I remember him imposing herbaceous discipline upon the unruly expanse of the local church, back before it was engulfed by an old people's village. There he'd be, happily swearing away, safe in the knowledge that the roar of the machine drowned out his invective in the ears of fellow parishioners, if not in God's. One of my early memories, from the age of three or four, is of being given a toy mower. It made a most satisfying clackety-clack as I pushed it around, and I'd follow in Dad's footsteps as he shaved another half-centimetre off the top of the front lawn. But somewhere between preschool and adulthood, I lost the urge to mow. It was Dad who gave me the Rover as he upgraded to a newer, brighter model, perhaps hoping home ownership would belatedly convert me to the cause.

'Plenty of life left in this one, mate. Briggs and Stratton motor. Yanks used 'em in the war.'

Yet now, as I give the lawn its annual mauling, I realise an era is spluttering to a close. The chorus of the suburban weekend, of a dozen Victas singing in discordant harmony, has been largely silenced. I finish the backyard, shut down the engine and listen. Nothing. Not another to be heard. Like the last of the dinosaurs, doomed by climate change, the Rover has sent out its plaintive call but has received none in reply.

Of course, it could just be affluence. Perhaps the weekend cacophony has simply been outsourced to the operatives of Jim's Mowing, slipping in under cover of office hours with their stealthy, efficient machines. But I don't think so. Up and down our street, lawns once sprinkler verdant and blood-and-bone luxuriant have grown brown and patchy under mandatory water restrictions. Yesteryear's heresy has become today's badge of civic responsibility. Some friends have taken their lawn out altogether, replacing it with woodchips, mulch and drought-resistant natives. Water tanks, once banned as an eyesore in the national capital, have been installed to keep the wattles, gums and bottlebrushes alive. They've put in a small patch of astroturf to give the kids something to roll about on, like a hands-on display at the natural history museum.

I fossick under the house to see if I can find a sprinkler. Eventually I locate one, weathered and abandoned, together with a disused hose. I don't much miss the sound of lawnmowers, but I fondly recall the sibilant swish of the sprinklers. It's a soundtrack from childhood: drifting off to sleep on a summer evening with their tack-tack-tack floating through an open window. My kids, aged six and ten, have never known sprinklers or the joy of running under them in the dry heat of summer. Denied their birthright, they have had to make do with 'car wash' instead. In Canberra, even washing a car has become prohibited. So in warm weather, on the occasion of a rare cloudburst, we yell 'Car wash!' and the kids flap around deliriously, trying to find long-neglected raincoats while my wife and I put our cars out to catch the downpour. The whole family goes at them in a lather of bubbles and laughter, scrubbing away and hoping the rain lasts long enough to rinse them off.

But now, mowing duties fulfilled for another year, I put the Rover back to sleep, pack the car and prepare to depart, contemplating the garden as I do so. It's not just the grass. We've lost two large gums to drought, and a massive ash tree in the backyard looks like going the same way. Its shallow root system, established in a time of plenty, doesn't tap far enough down to reach the receding water table. I start the car, thinking I'm off to find out what's happening to Australia's rivers, to some other place. But I realise, like the frog in the saucepan, that I've been sitting here blissfully unaware while the temperature has been rising all around me. The crisis is here and now.

It's early November as I head towards Queensland and the headwaters of the Darling. I've loaded up my ageing Korean station wagon with all manner of stuff: camping gear, food and jerry cans are jostling with my camera, notebook and voice recorder. Rather than planning out my needs clearly, I've stuffed in pretty much anything that will fit, including an Esky, a gas stove, even my old guitar. They rattle away companionably as I make my way past Parliament House, its flag limp in the late spring heat, and head for the open road. I'm travelling north, an 800-kilometre drive to Moree where I'll pick up a four-wheel drive and go bush. For the first few hours I track through familiar territory—up the highway through the vineyards of retired bureaucrats towards Yass, then west for a short way along the high-speed anonymity of the Hume, and then north again along the Lachlan Valley Way, a winding secondary road. There was a time when I drove this route frequently, back and forth between the New South Wales central west and Canberra during the three years I went to uni at Bathurst. Back then, in the early 1980s, it was a forsaken stretch of land, hammered into submission by an intense drought. I drove it once with a friend, a farmer's daughter, and we stopped to watch emaciated sheep staggering through a paddock completely devoid of grass, looking for all the world as if they were trying to eat rocks. But now, as I speed through lush dales and over well-grassed hills, the countryside exhilarates me. It's ripe and overgrown, with a persistent tint of late-spring green. It's as good as I've ever seen it. The Murray-Darling is running on empty, the lower lakes are at tipping point, and the drought persists. But not here, not now.

After an hour and a half I reach Boorowa. I'm not planning to stop, but I can't help myself. Back in my uni days, this had been a town on the edge. I'd stop at its one tired cafe for a toasted sandwich and a cup of dishwater coffee, occasionally fossicking through the neighbouring bric-a-brac shop on the rare occasion it was open. Now

I discover a Boorowa with five cafes as well as that irrefutable seal of rural viability, a Chinese restaurant. 'Hurrah, Hurrah for Boorowa, The centre of Australia-R' reads a sign on the way into the metropolis of 1200.

I sip a macchiato at a brand new cafe decked out with blonde-wood furniture and terracotta floor tiles, commenting to the young barista how green the countryside is looking.

'Wouldn't know mate', he says. 'I don't get out of town much.'

It's not just Boorowa that's changed. There was a time when country towns kept to themselves, giving little away to an errant motorist. The signs on the highway would recite, prisoner-of-war style, only the town's name, population and altitude: 'Walgett: population 2300, altitude 130 metres'; 'Collarenebri: population 1234, elevation 150 metres'; 'Mungindi: population 1000, elevation 160 metres'. Why anyone would want to know the altitude of a town marooned hundreds of kilometres from the nearest mountain, I have no idea. The average annual rainfall would be more useful. But then again, elevation is unchanging and unchallengeable; who can say what the average rainfall is anymore, or even what it means?

These days, however, towns are no longer satisfied with Geneva Convention minima. Now every town requires a vision statement: 'Boorowa: Superb Parrot, Superb Country'; 'Cowra: World Centre of Friendship'; 'Gilgandra: Town of Windmills and Home of the Coo-ees'; 'Narrabri: The Centre of Agriculture'. Later in the day I'll stop in Belatta, perplexed by a sign reading 'Belatta: Between the Mountains and the Plains'. What mountains? You mean those low grey shapes over there on the horizon? Still, it's perhaps more plausible than the 'World Centre of Friendship'.

North of Boorowa, the country continues to display remarkable fecundity. The other side of Dubbo I pass a wheatfield from central casting, a sea of golden grain rippling in the breeze. Off to one side stands a fibro farmhouse, the crop lapping almost to its door. I imagine children inside. After seven long years of drought, I conjure up a Christmas to remember, lavish presents compensating for years of doing without. It's one of the great pleasures of long-distance driving: daydreaming away the passing miles.

Stopping for lunch, I check in with reality, the reality of official reports and statistics. I revisit this week's drought update from the Murray-Darling Basin Commission. It spells out a crisis for the southern part of the river system, down in Victoria and South Australia, with record low rainfall, record low inflows into the rivers, and climbing maximum temperatures. But it's more ambiguous about the north, up where I'm heading:

> In the northern half of the Basin, there has been some good rainfall over the last few months and this has assisted dryland farmers. However, the rainfall has not been sufficient to pro-duce significant runoff and streamflows in the Darling and its tributaries have generally remained very low.

I look at the wheatfields, glowing gold in the afternoon light. Everywhere I look, the country speaks of water and wealth. I don't get it. Whatever is happening down in the south, up here the rain has been falling. So how come there's no water in the rivers?

At Narrabri, overnight rain has left puddles beside the road and a few stagnant pools in Narrabri Creek, but nothing at all in the Namoi River. Indeed, the recent downpours have caused an explosion of growth in the riverbed, and the town council has fastidiously mown it for 50 metres on either side of the bridge I drive across. Then again, Narrabri is cotton country. Glimpsed from the road, the 8-metre-high walls of giant on-farm water storages suggest why there is no water left in the river. At Collarenebri, agriculture has emptied the Gwydir, yet the nearby Barwon is flowing.

The land stretching north-east from Bourke tells a different story. It's flat here. Flat, dry and hot. Sitting under the Queensland border 800 kilometres inland, it doesn't receive enough annual rainfall to even consider planting a crop like wheat. This is grazing land, where an 8000-hectare property is marginal and 2 hectares or more are required to support each head of sheep or cattle. I stop where the dead-straight dirt road bisects a giant claypan shimmering in the late-afternoon heat. Spring has another month to run but the temperature has climbed to 42 degrees. The air is oven dry; it tugs at your skin, trying to suck out the moisture. Off to one side, the relic of a car bears testimony to some long-ago drama. I walk over to inspect it, carrying the weight of the sun on my back. The wreck has the accumulated patina of decades; there is nothing left save metal, but little rust. I climb onto its roof and look around. Nowhere is there a hill, or even a mound. On the gleaming horizon I can see the curvature of the earth, out where the cloudless blue of the sky is bleeding down into the plain. To the north and east there's a straggle of mulga. All around me there is complete silence. There's no sign of animal life. The air smells of nothing but heat.

I've left my city car back in Moree, 400 kilometres to the east, where I hired a four-wheel drive to cope with the variable roads of the outback. Back in the car, with the air conditioning cranked up and the iPod feeding the stereo, the speedo reckons I'm flying along at 110 kilometres per hour, but the featureless countryside appears to be passing in slow motion. Almost subliminally I start to perceive slight alterations to the land passing beneath the wheels: for a kilometre or two there is red-soil country, covered in tiny pebbles or russet sand, then black-soil country, with grey sand so fine it's like talcum, and then red soil again, interspersed with the occasional claypan or sprinkling of whitish sand covering the underlying rock. The only high points are cattle grids every 5 to 10 kilometres. The black-soil country, I realise, is the flood plain—the ash-like soil is the silt deposited over centuries by water moving across this moonscape at times of flood—whereas the red-soil country is the high ground. I stop the car again at a border between the two types of soil and walk away from the road's slight elevation to glean the subtle difference in the landscape that sends water one way instead of another, but I can't. Only water and gravity, conspiring to find the lowest point, can discover the way. Out here, the smallest man-made alteration—a fence line, a road, maybe even the wheel marks of farm machinery—has the potential to send floodwaters sprawling in an entirely different direction.

Belatedly, I realise I have already arrived at my intended destination: the flood plain of the Culgoa River. There has been no sign—

'Culgoa Flood Plain: population declining, altitude unchanging'—
no vision statement. I search the horizon, but the intermittent lines
of mulga give no clue as to where any permanent watercourse may
lie. And yet floodwaters here can spread for 40 kilometres. Another
realisation approaches me through the oppressive treacle of the heat.
All those town elevations I've been sniggering at do mean something
after all. Instead of noting vision statements, I should have been
paying more attention to the heights above sea level. In a land so flat, a
fraction of a metre can decide the flow of water and all that goes with
it. Dirranbandi, Queensland, near where the flood plain starts, has an
elevation of 172 metres. Goodooga, 100 kilometres downstream, is
at 142 metres. Bourke, another 200 kilometres as the crow flies, and
maybe 600 river kilometres, lies at 106 metres. From there it's only
100 metres of fall, the length of an up-ended football field, to carry
the water all the way to the Southern Ocean.

Intrigued, I check my maps. Sure enough, they are unanimous
when it comes to the alignment of roads, in complete agreement on
the location of towns, but vague and non-committal on the location
of the river I've come so far to see. Some show it flowing into the
Darling; one has it pulling up well short. On one it is a well-defined
and confident line; on another it winds this way and that in a drunken
squiggle; on yet another it's not a single strand but a weave of blue
hairlines. The river, it seems, has confounded the mapmakers. I suspect
that, like myself, many of the cartographers have made the mistake of
having a preconceived idea of what the word 'river' means. Perhaps
they looked it up in the *Oxford Dictionary*: 'Copious stream of water
flowing in channel to sea …'

The Darling, by some measures, is Australia's longest river, and
one of its most significant. Yet in the same spirit of antipodean
perversity that presented the colonising English with the platypus, it
defies preconceptions of how a major river should properly behave.
Whereas the Amazon, the Ganges and the Nile place their deltas neatly
down in the lowlands near their mouths, the Darling has decided to
insert its delta up beyond where it officially begins. The Condamine-
Balonne rises traditionally enough near the Queensland towns of
Warwick and Toowoomba on the western slopes of the Great Dividing

Range, gathering in tributaries in the time-honoured way as it heads towards Dirranbandi, nearly 400 kilometres to the west. But after Dirranbandi, the flatness of the earth splits the Condamine-Balonne apart. It becomes the Culgoa, the Birrie, the Bokhan and the Narran. So flat is the land that the Narran gives up flowing altogether, ending instead in the terminal depression of the Narran Lakes. The other three rivers track this way and that through this delta of interchanging channels, criss-crossing their interlinked flood plains before eventually joining the Barwon. Indeed, where the Culgoa joins the Barwon is the official beginning of the Darling River. That's the official version, the book version. Yet the one map I have that shows the Culgoa stopping well short of the Barwon is right more often than not. It certainly is this year, despite the recent rains. For by the time the main channel of the Culgoa reaches the Barwon, there is not a single drop of water left in it. A 'copious stream of water flowing in channel …'? Give me a break. The start of the Darling has become no more than an arbitrary point on a map.

The Fennels have been at Boneda since 1883, the property's 8000 hectares carrying 4000 sheep on the western bank of the Culgoa. It's hard country. The red and black soils struggle to support outbreaks of lignum and other native shrubs between clumps of mulga and gidgee-gidgee trees, even in this, a good year. Here, properties are not handed down from father to son with the easy assumptions of prosperity. Instead, eldest sons borrow from the bank to raise the capital to buy out their own fathers. That's what Hugh Fennel did back in 1968, and that's what his father did before him. But Hugh and his wife, Pam, have no sons. Their two daughters, both professionals, have long ago left the district: one for Brisbane, the other for Dubbo. Boneda has been intermittently on the market for more than a year. Soon the Fennels will go to Dubbo as well.

'I'm planning on doing absolutely nothing. I'm just going to read and go for walks and [do] whatever it is that retired people are supposed to do', Hugh laughs. 'I'm sixty-four now and I'm working harder than I've ever done.'

Theirs is a modest house, neat and comfortable, set back 50 metres from the Culgoa's main channel on land imperceptibly higher than its surrounds. The green of the garden, enclosed by a high fence to keep out kangaroos and goats, clashes with the bare red soil of the surrounding country. Inside there are none of the trappings of the squattocracy, just simple furniture and an old-fashioned television carrying the ABC news, keeping the couple abreast of events from the wider world, what Pam refers to simply as 'away'. 'You wouldn't understand', she says at one point. 'You're from away.' A large painting hangs on the living room wall, depicting a daughter's wedding. Over a couple of cold beers, the Fennels tell me there's every chance the house will remain empty once they sell.

'I don't think we'll be selling locally because most of our neighbours are thinking of doing the same thing—getting out', says Pam. If a local was to buy, then it would be for the land, not the house; the same for an absentee landlord. Hugh would like to sell to the National Parks and Wildlife Service, for he still cares for the land his family has tended for well over a century: 'National Parks look after the country. It would never be stocked for a start. It would just look the way it's supposed to look, and that would be a good thing. If a southern buyer got hold of it and overstocked it, it could turn into a wasteland. It's taken a hell of a hammering over the years. It's just a fluke we got the 9 inches late last year and the 8 inches in January and February this year. It was just a godsend. There is no other way to describe it'. I think of the desolate claypan out by the wrecked car and my imagination fails me: how much worse could the country have been before the rains?

When the Fennels go, they won't be the first, or the last. Once, every property in the district employed a married couple, as well as stockmen and local shearers. But no more. The Fennels remember when the closest town, Brewarrina, 80 kilometres to the south, hosted a dance every Friday and Saturday night and a ball five or six times a year. It used to have 3500 people and six cricket teams. Now there are 800 inhabitants and there hasn't been a cricket team for a decade.

'The work is getting too hard. There are no stockmen, no shearers. There are only a quarter of the people around here that there used to be. There are no young people. I still consider myself one of the young ones, but I'm having myself on', says Hugh.

That night I sleep out in the bunkhouse and think of all the happiness and heartbreak Boneda must have witnessed over the years: of the long-gone daughters riding their ponies; of Hugh's father selling his son his own birthright; and of the river, 50 metres away, that has done nothing to help in the struggle. Hugh tells me that he depends almost entirely on rainfall in a country where rainfall is rare and erratic: 'What comes down the river is enough to look after the garden and that's all. We don't even use the river for stock'.

The next day, Hugh is off to work before I'm up, rising early to beat the heat, but Pam walks me down to look at the stream. It's a turbid yellow and isn't flowing. She explains that it's full of boron, which is slowly killing the garden. The water hasn't come down the Culgoa proper but down one of those here-again-gone-again squiggles on the map called the Nebine Creek, before settling in the Fennel's waterhole. At least it's water.

⌐⟶

I drive back out of Boneda, across the flood plain that no longer floods, again heading north. An impressive bridge takes me across the empty Burban Creek and I come to Woolhara. It's not the first property I've come across that evokes the suburbs of 'away'; I've already passed a Kirribilli and a Bronte—such strange names, stranded out here hundreds of kilometres from the ocean. At first glance Woolhara appears similar to Boneda, with the same low-slung house surrounded by a high homestead fence, and a windmill slowly turning. But when I stop, there are no yelps from working dogs and the lawn has gone to dirt and weeds. Water from the windmill hose leaks unminded, gathering in a pool beside the fence. There is no human sound, just the wind, the drip of the water, and the groans of the steel machinery shed expanding and contracting in the sun. Firewood, neatly chopped and stacked, rests against an empty kennel. A water gauge beyond the homestead fence shows 40 points, although it hasn't rained for days. How long the house has been abandoned I can't tell, but it must only be six months or a year. I consider pushing through the unlocked gate and trying the doors, but some sense of trespass holds me back.

Instead I walk back across the drive and scramble up the wall of a small dam. It holds plenty of water but there is no sign of stock. Not far away lies a well-tended grave, its galvanised iron fence shining rust-free in the shade of a small gum. The concrete slab is clear and in good repair, and the inscription on the marble headstone shows little sign of erosion. It looks recent, but it's not.

> In loving remembrance of my beloved mother, Elizabeth
> Popplewell, who died May 6th 1899 aged 69 years.
> Sleep on beloved sleep and take thy rest
> Lay down thy head upon thy saviour's breast
> We love thee well but Jesus loves thee best
> Good Night.

A passing emu has left its three-pronged footprints set deep into the red clay outside the fence like the tracks of some long-extinct dinosaur.

I drive through Woolhara and across a private bridge made of steel, which arches gracefully up and over the main channel of the Culgoa. The bridge is barely wide enough for the car; it looks and sounds more like a footbridge, its red-metal sheeting ringing under the tyres. Below its arch, the bed of the Culgoa is bone dry. A young grazier called Alan Thompson built the span back in the 1950s. It's Alan, now eighty-three, whom I've come looking for.

The track winds through the eucalypts of the black-soil flood plain, past the Thompson's Caringle homestead, before straightening up and widening into the red-gibber plain of an airstrip. A plane is nestled in a corrugated iron shed. I floor the four-wheel drive, sending emus hurtling out of the way as the car screams down the runway. The landing strip narrows back into a track, I slow down, and the emus stare disapprovingly from the distance.

Alan is working with his son David and David's wife, Belinda, in the mustering yards of the neighbouring property, Boolaboo. Into his ninth decade, he's still lean and erect, and displays little of the stiffness of back so common among farmers. We escape the mounting heat for tea at his son's homestead. The two properties, Caringle and Boolaboo, are run jointly: 22 000 hectares with about 16 kilometres of river frontage.

'When I bought Caringle in 1952 the river was pretty well a constant stream', Alan tells me. 'It would run for at least nine months of the year and sometimes it would run the whole year. It was considered an asset to have the river as a watering point. Now it's become a liability to have the river, because it's dry most of the time. We can no longer depend on the waterhole in the river because it's silting up. We've had to spend quite a bit of money fencing the river off ... Sheep get into it and they're hard to get out ... We used to see quite a few turtles and river rats—they're not rats really, more like a native otter ... but we don't see them anymore because the waterholes have all dried up. There's been nowhere for them to go. The same with the old mud turtles. I haven't seen one of them for years.'

The old grazier tells me that the river used to flood three or four times a year, covering 8000 to 12 000 hectares at a time. But there hasn't been a flood for ten years and the river is dry most of the time. As Alan talks about floods, I begin to reconsider the concept, just as I've reassessed my concept of a river. A flood in the outback, with its wide flat spaces, is very different from a flood down on the coast. On the coast, a flood is a sudden, violent thing, coming without warning, breaking riverbanks, threatening lives and sweeping away homes and bridges. Out here they are mostly gradual things, coming with many weeks notice as the water eases its way down from Queensland. Slowly, the water moves out across the plain, and slowly it recedes. And rather than wreaking havoc, the flood brings silt, topsoil and nutrients to replenish the parched land. Graziers have been known to buy additional stock in the time between the flood rising in the north and its arrival at their properties.

'We've been depending on the rainfall', says Alan. 'The land's still responding but it needs a flood every now and then. They're a bloody nuisance when you've got them. They're good afterwards. The '56 flood was the biggest flood we had here. It rained cats and dogs from October '55 to July '56. We had about four floods. We had a bloke come here wool classing and he left his vehicle here. I took him up to another place over the border. And it rained and rained and rained, and his vehicle was here for nine months before he could shift it … In 1981 there was a big flood in Dalby, up in Queensland, and we got a big flood out of that. And when the flood went we had Dalby topsoil everywhere. It was pink; it turned the black country pink for a while. Dalby had lost about 18 inches of topsoil. They were crook something awful about that.'

There was a time when Alan and his wife Elaine employed stockmen, jackeroos and a full-time cook. Each property in the district housed its own small community, many of them local Aboriginals. But drought and the slow death of the river, combined with falling wool prices and rising labour costs, have meant that the only hired help employed nowadays are seasonal shearers and crutchers.

'The bloke we had here had eleven kids', Alan recalls. 'At Woolhara they had a bloke with six kids. So I used to put them in the back of a Bedford tip truck to take them to school, and if they started fighting, I'd start lifting it up, so they had to stop fighting and hang on.'

I ask Alan what happened to Woolhara. He tells me it's owned by an estate, originally bequeathed in the 1930s, which has put it up for sale.

'They had a manager there. The manager's wife had cancer and she eventually passed away. And the husband, he was a very fastidious bloke. He got everything in order on the place, neat and tidy and shipshape. And then he sat in his car and gassed himself. And there's been no-one living there since. It happened about two years ago. They were very good neighbours. Very good.'

⌣

I find Fred Hooper reclining on the couch in the once grand sitting room at Weilmoringle, twenty minutes up the road from Alan Thompson's Caringle. Fred is wearing a singlet and boxer shorts, and is transfixed by the 152-centimetre plasma television that dominates the room. On the screen, Rowan Atkinson's rubber face, twice life-size, is twisting itself into ever more unlikely contortions as Mr Bean struggles to make himself understood by a pretty French girl. Fred gives one last chuckle. 'Sorry, mate', he says, flicking off the telly and extending his hand. I take a seat in a worn armchair next to the couch, the room dim now that it's been deprived of the wide-screen's glow. After the blazing midday light outside, my eyes takes a while to adjust. I can make out an impressive fireplace, and walls lined with cypress pine.

At its peak, Weilmoringle was more like a principality than a property, stretching half a million acres across the Culgoa flood plain and beyond. The homestead, built in 1883 and lying 120 kilometres north of Brewarrina, was part home and part town hall, the centre of a grazing empire. But Weilmoringle wasn't built on virgin ground. The house and its outbuildings sit next to one of the few permanent, or near permanent, waterholes on the Culgoa, which for time immemorial had been a camp of the local Morowari people. And so the squatters settled close to two essential commodities: water and labour. For while the Morowari lost their land, they remained living upon it and were offered employment instead. It was a poor trade: the work was hard and the pay low, but at least the community retained its identity.

'These big properties relied on the Aboriginal labour', Fred tells me. 'Without that Aboriginal labour some of these big properties probably wouldn't have survived. So they made them sanctuaries to stop people being taken off and put onto missions. I know one other property, over on the Warrego, that was like a sanctuary as well. The station owners wouldn't let the protectors come on and take Aboriginal people off.'

And so while other Aboriginal communities fractured under the policies of assimilation, of missions and boarding schools, the community at Weilmoringle endured. The Morowari worked on the property as stockmen, shearers, boundary riders and domestic workers. Others also lived on the property, enjoying its protection while working as casual labourers on the small holdings of the district. Still others lived and worked full-time on surrounding properties while maintaining family links at Weilmoringle. Life was hard, but not unbearable. But with the advent of equal pay for Aboriginal workers and the collapse of the wool economy, the coming of drought years and the changes in the river, the small holdings could no longer afford to employ full-time workers. Weilmoringle itself was carved up into smaller holdings. Many younger Morowari drifted away to lose themselves in the grog and gambling of Brewarrina and Bourke. Nevertheless, about sixty Morowari still live in the Weilmoringle community, within a dozen or so houses gathered across the road from the homestead, next to the river.

Fred Hooper is Morowari. He spent the first decade of his life, in the 1960s, growing up in the shadow of the homestead—playing in the river, swimming, building boats from corrugated tin, fishing. 'We'd be in the river the whole time', Fred recalls. 'We relied on the river. I remember Mum used to go down with the old yokes with the 20-litre drums and bring them up and put them in the 44-gallon drums. The sediment would drop to the bottom and the water would be clear as crystal.'

Now the Indigenous Land Corporation has bought the property, the Morowari have reclaimed Weilmoringle, and Fred has moved into the homestead. 'When I first came to this house I couldn't sleep, because traditionally we weren't allowed inside the yard. In the old days, when you came to collect your wage, you had to wait outside for it to be brought out to you. Very few Aboriginal people would have sat inside this house', he says.

Fred is beginning to warm to our conversation. He's a gregarious bloke in his late forties, with a ready laugh and a mind full of insights. He spent his teen years in Wee Waa, had six years in the Navy, and has worked around the country for various public service agencies. He's returned from Tasmania with his partner, Kylie, to look after Weilmoringle on a voluntary basis. He reckons race relations in the district are good: 'Really good. You get some old rednecks. Well, not rednecks. You get people with old traditions and old beliefs; not necessarily rednecks, but conservative. But generally, it's good here'.

Back in the days of empire—British and grazing—the homesteads of the squattocracy were the social as well as the economic hub of the far-flung grazing districts. That tradition continues; the old property still hosts social events for district landholders. 'When we have tennis days and community days up here, it's mainly the white property owners around the area come up, but you get a couple of people from the community come up as well', says Kylie.

Fred and Kylie would like to continue the grazing on the property—it's currently carrying 200 cattle and 5000 sheep—and there are plans for tourism based on Weilmoringle's unique bicultural past. Offering employment to the local community is a priority. So, too, is protecting Aboriginal sites, and there are hopes of restoring some of the district's biodiversity. It will be no easy task. Reduced to just 17 000 hectares, Weilmoringle is no longer the vast fiefdom it once was. The ILC has bought another 8000 hectares on a nearby property. Fred is hoping more properties can be added to create the necessary economies of scale. It's an ambitious plan, hatched at a time when even the most experienced graziers are finding it hard to break even. As one of them told me, running a property by committee is not a recipe for success.

But Fred is optimistic. 'I think we can bring it back—if we look at a balance between conservation and business. And respect the capacity of the country. In terms of providing jobs, the bigger we are, the more jobs we can provide', he says, before looking at his hands and lowering his voice. 'And where the people want to work. That's another big factor. Aboriginal people have relied on the welfare system for too long. I think it's time the wind changed, that we ourselves accept the fact that we live in a white society and we need to make the best of that society. It's about getting out, getting up off your arses and going out and trying to create a job.'

Fred and Kylie give me a tour of the old house, along its wide central corridor lined with glowing cypress pine, into a wood-lined attic that is beginning to swelter from the assault of the sun, and up to the widow's walk that extends along the ridge of the roof. The land all around is flat and dry. In a land without hills, the view reaches into the never-never, the scrub and the grass a uniform dun. There's little wind, but something has lifted a thin miasma of dust into the air so that the blue sky towards the horizon is tinted brown.

Fred and Kylie take me to see the old shearing shed, built in the 1890s, empty now, but which once housed 100 stands. There's an old steam engine, like a disembodied locomotive, its belts stretching off towards cranks and wheels that powered the shears. They're not sure how long it's been since the shed has been used, but the smells still linger: sweat, manure, wool. The floorboards are smooth, soaked through with lanolin and polished by years of work boots.

Back at the house, we go out the back gate and down to the river. There's water in the waterhole, but it's brown and low and not flowing. It ends about 50 metres downstream from our vantage point, collecting behind a natural weir. Fred won't swim in the waterhole now, not like when he was a kid, not since he had to fish a dead kangaroo out. He says it reminded him of the stories the old people used to tell—of bad spirits lurking in the deep water, waiting to prey on naughty children. 'The river is dying, and because the river is dying, the community is dying as well', he says. 'We were living off the river … because the river was flowing. Now the river isn't flowing, the fish aren't there … The baby mussels are dying out. There were baby mussels everywhere. We used to get mussels this big. You know, 7 or 8 inches long. Yeah, we used to get them like that. I remember as a kid, this river never stopped flowing. Look at it now.'

⌒

I meet up with National Parks ranger Bart Schiebaan at the Weilmoringle post office. He's come out from Bourke for a day or so and has agreed to show me around the Culgoa National Park, a collection of old grazing properties up by the river on the Queensland border. Bart is a strapping young bloke in his twenties, well over 6 feet tall and bristling with muscles and enthusiasm. Today, he's also on a tight schedule. He leaps into his Parks truck and I follow him north-west out into red-soil country. I'm doing between 100 and 110 kilometres on the dirt, but my city skills are no match for Bart. In no time flat he's just a ball of dust on the horizon. Some minutes later I catch up as he waits at a turn-off. We repeat the exercise as he again leaves me far behind before waiting at another turn-off, this time leading into the national park. We proceed more slowly now, winding along an infrequently used track that carves into black-soil country. Around me there is a fair amount of growth: stunted trees and intermittent ground cover. But to an untrained eye the national park doesn't look in any better condition than the grazing country down at Boneda, Caringle and Boolaboo.

We finally reach a couple of desolate looking picnic tables failing to find shade under a ragged bunch of black box and coolibah trees. Bart gets out wearing a look of concern. He tells me the picnic and camping ground has been deliberately set back 100 metres or so from the river, back off the flood plain. It all looks dead flat to me. We walk along a sad little trail to the river as he points at various types of trees, bushes and grasses. My eyes haven't deceived me: Bart confirms that after a decade without a flood, many of them are dying.

'This has all just suddenly died off in the last few years', says Bart. 'It was all okay back in 2000, because we put in the walking track here, and it was a beautiful little walk to the river, and then it all just decided to die. So it's not the most scenic walk for a visitor anymore. And these trees are great communities for bird life. This national park has twice the biodiversity of other parks out this way, but I suspect that statistic is taking a bit of a hammering.'

We reach the river. A lone picnic table looks out across desolation. The watercourse here is bigger than at Caringle, bigger than at Weilmoringle. The banks must be 30 metres apart and the bed of the river 10 metres deep. But it's empty. Off on one side, an Aboriginal scar tree, a river red gum that long ago had a shield cut from its bark, is collapsing down into the empty river. Most of its branches are grey and barren, though a few tufts of leaves sprout bravely from one. 'It'll be dead by next summer', Bart says matter-of-factly.

We scramble down the banks into the empty river. The bottom is dry grey moon dust; my boots leave an impression that could have been made by Buzz Aldrin. I scoop some of the powder up into my hand; there's not the slightest hint of moisture in it. Bart tells me there is a more or less permanent waterhole just around the bend and we talk as we walk to it.

'I don't know if it's to do with climate change or what; it's just getting bashed from every angle. Whether it's goats, or climate change, or irrigation. It's just getting hammered … We have koalas in the park. It's the western limit of where koalas are found. They can live without water by getting their moisture from the trees, but if the river red gums and other trees die … well …' He trails off. We've reached the waterhole.

Bart seems to be talking to himself now as much as he is to me. 'God, that's dry now too. There was a pool of water in there. And now I'm surprised to see—well, I guess I'm not—but it's empty. I was here about a month ago and there was about a foot of water in there. And it's now dry. You know, it hasn't actually been such a dry year. It's actually been quite a wet year. Shit. And now it's empty.'

We walk back to the cars in silence. Bart's enthusiasm has dried up like the river, and he's subdued as he climbs back in his truck and bids me farewell. He asks if I'm okay to find my own way back. 'No problem', I say.

Once he's gone, I retrace our steps along the parched little tourist trail down to the river, sit at the forlorn picnic table and stare into the bottom of the empty river. I'd been thinking of camping here, but it would feel like camping at Woolhara.

There once was a swagman who camped by a billabong
Under the shade of a Coolibah tree
And he sang as he looked at the old billy boiling
Who'll come a-waltzing Matilda with me.

But the swagmen are long gone, as are the stockmen and the jackeroos, leaving only a dwindling number of graziers. The billabongs have gone, too. How can there be a billabong when the waterholes in the main channel itself have dried up? The coolibahs are dying and there isn't enough water in the whole of Culgoa National Park to fill a billy. Only the ghosts remain: the ghosts of the swagmen, of the Aboriginals who cut the shield from a once majestic river red gum, and the ghost of the river itself.

⌒

St George is awash with water and money. Or so it seems to me, newly arrived from the desiccated banks of the Culgoa. The town lies beside the Balonne in Queensland, ostensibly the same river but about 200 kilometres upstream. I sit beside St George's quiet lagoon in its late-afternoon glory, watching the ducks floating by and the cranes wading nonchalantly through the rushes at the water's edge. A pelican has picked up a thermal and glides effortlessly overhead. The river here is 50 metres wide, more like a lake than a stream. The water is brown and still, its surface rippling as it's brushed by a breeze that comes off the water cool and refreshing. The crowns of long-drowned red gums break the surface mid-river, remnants from the time before the town's weir was built, in 1972. At least, it's called a weir—'The Jack Taylor Weir', states a large and shiny sign—but it looks more like a dam to me. It's solid concrete, 6 metres high, with thirteen sluice gates and a highway running across the top of it. From the weir/dam the water backs up the length of St George and for many kilometres beyond, almost as far as the larger Beardmore Dam, about 20 kilometres to the north-east, completed in the early 1970s. On the town's outskirts, between it and the dam, a kind of millionaires' row overlooks the river. Designer houses sprawl on single-hectare blocks, mounted by evaporative air conditioners and satellite dishes, a hobby vineyard here, a carpet of lawn there, a shiny four-wheel drive parked on a gravel drive. The backyard pergolas face the sun setting over permanent water, a view guaranteed by the two dams: one upstream, the other downstream. Two hundred kilometres from Culgoa National Park. Two hundred kilometres and a world away.

A powerboat erupts into life 100 metres down the lake, breaking my reverie, and I walk down to investigate. The engine roars again and the white fibreglass dart accelerates away, pulling two waterskiers behind it. They're both kneeling, attempting to weave back and forth across the wake. One falls, then the other, and the boat throttles back and eases around to collect the skiers as they laugh and tread water. On the shore, two young blokes, strutting bare-chested and board-shorted, their beers enshrined in Fourex stubbie holders, have put another boat in the water and are collecting fishing rods from the back of their truck. A teenage girl in a black bikini offers a shy smile from the back of the boat.

'Good day for it', I offer.

''ken oath, mate.'

We chat as they finish loading the boat and then they're off, heading upstream away from the town, chasing yellow-bellies. I continue along the path through the Blondie Corrington Riverside

Reserve, the long narrow park that stretches for more than a kilometre between the town and the river. I'm still having trouble coming to terms with all the water. A battery of sprinklers, chattering away among themselves in the heat of the dying day, is keeping the well-trimmed lawn a pleasing green. The town is on Level 1 water restrictions, the least onerous level. A council worker trundles past on a ride-on mower, a miniature John Deere.

The Blondie Corrington Riverside Reserve has been divided evenly between three service clubs: Rotary controls the stretch closest to the weir, Apex has taken the centre ground, and Lions has bunkered down in the final third. None are about to concede anything to their rivals; the park is immaculate, with picnic tables, a couple of artificial beaches created from trucked-in sand, and a plethora of public toilets—there are two in the reserve itself and at least another three placed at strategic intervals across the road. I walk beside the river along a concrete footpath linking the three sections. Imprinted in the concrete at intervals of 30 or 40 metres is the word 'cotton' accompanied by a stylised cotton puff, looking rather like the clubs emblem from a deck of cards minus the stalk; a constant reminder of what has ultimately paid for the languor of the park, its abundance of toilets, and its sense of achievement.

The reserve is bristling with plaques. At least three commemorate the day in 1846, St George's Day, when explorer Sir Thomas Mitchell camped here and crossed the Balonne. Another marks the location of a time capsule buried on the 150th anniversary of his crossing. There's a plaque marking the shire's centenary, in 2002, another dated the same year acknowledging World Peace Day, and a couple more 'commemorating' work done on the park by the Department of Corrective Services. There are plaques remembering the fallen of World War I, World War II and (collectively) all the other wars, plus a fourth honouring those who erected the first three; another plaque dated 2005 gets in on the act by re-dedicating the first four. There are plaques acknowledging the World War II exploits of two local pilots, one of them the only Aboriginal pilot in the entire conflict, Leonard Waters, who flew a Kittyhawk fighter with 'Black Magic' painted on its nose. Yet another plaque names the politician who unveiled the pilots' plaques. There's also a plaque marking the spot of another time capsule, buried in 2005 by a veteran of Thailand's Hellfire Pass, and due to be opened in 2055. So many plaques. Here is a town that believes itself worth remembering.

I walk across the road into St George itself. I pass the Asian Pearl restaurant, where a demure waitress flips the sign behind the door from 'closed' to 'open'; 'Chinese and Australian Meals' reads the sign in the window. Other businesses have closed down for the night: the newsagent, the cafes, the bakery, real estate agents, accountants, solicitors, a dentist, an optometrist. A street-sweeper drives past on its way back to the depot. A big supermarket is still open, its car park half-full, but the folks at Boats, Bullets, and Blades have called it a day. Even after seven years of drought, this town of 2500 remains resiliently prosperous. There are no boarded-up shops, few if any 'for lease' signs. A Toyota dealership displays a yard full of shiny new vehicles: there are plenty of sedans and a lower ratio of four-wheel drives than in some Sydney suburbs. St George is a 500-kilometre drive west of Brisbane, but it feels much further east.

It's the end of a long day. I've spent it driving up via a circuitous route through Moree after reclaiming my car. As I approached St George it was difficult to believe it was even remotely in the same part of the world as the sun-baked flood plains of the Culgoa. North of the border and a couple of hundred kilometres further east the rainfall is higher, maybe 45 centimetres instead of 30 centimetres. The difference means that dry land cropping is viable, or at least a gamble worth taking. This year that gamble has paid off big time. As I drove north-west, I encountered more and more wheat trucks, B-doubles, their trailers engorged with grain. Once, twice, three times, four, I pulled over to let oncoming headers pass, their massive cabs towering above the road like the heads of gargantuan insects, their harvesters towed behind like thoraxes. I watched the giant machines as they rumbled from fields devoured to those awaiting ingestion. Green headers, red ones, yellow ones—the primary colours of agriculture. Moving again, I'd caught up with a wheat truck, grain leaking in a steady stream from its rear door, until I found myself caught in a hailstorm of kernels, pinging as they ricocheted from the bonnet, clattering as they struck the windscreen. Wipers going, I took my chance, overtook and waved the driver down, bringing the juggernaut to a halt. I yelled to him of his slowly dissipating cargo.

'It's okay mate! Missing a seal!' he yelled back over the sound of his engine.

'You're losing a lot!'

'No problem! Plenty more where that came from!'

At Thallon I saw what he meant. The silos were filled beyond capacity. Next to them stood a makeshift mountain of wheat the size of a football field, piled three storeys high and partially covered in blue plastic sheeting, and another of similar size was almost complete; bulldozers were clearing the ground for a third. After seven years of drought, this will be a record harvest. World wheat prices are high. The drought here has broken, or at least gone into remission.

There are four pubs in St George: three across the road from the river, the fourth two blocks inland on the highway. It's a Friday night so I wander from one to another to see what this means in St George. There's a young crowd inside the art-deco Australian Hotel—lots of laughter and chiacking; blokes in one corner, girls in the other, with sporadic bursts of flirtation in-between. There are a couple of families at the St George watching the cricket on a wide-screen TV, and no-one save a couple playing pool at the Cobb and Co. At the Riverview about a dozen older blokes, farm labourers and council workers caked in the grime and sweat of the day's work, are holding up the bar. I have a couple of beers and a long, mostly incomprehensible discussion with a wizened bloke in his late fifties who seems to be called Raymond Hooper. I ask if he knows Fred at Weilmoringle but he shakes his head. The conversation revolves around brawling in Moree over the years and how St George is generally a better place to be. People are friendlier, less judgemental. And there's plenty of work. Raymond is positive I must have heard of his younger brother, a boxer. But entire slices of what Raymond is saying are indecipherable. He seems to have taken one punch too many in Moree, plus he's had a few drinks, and he has no teeth left except for two tiny brown pegs up front. The combined effect is to give his speech a wet, flapping sound, like a seal with a mouthful of fish. I decide the Riverview will do me and book a room out the back.

～

Chad Prescott drives a spade into the ground in a furrow between two rows of his cotton crop and turns the black soil over. Crouching down, he scoops it up into his hand.

'Here, smell this.'

I take the moist sod, rub it between my fingers and breathe in the earthy smell. There's the familiar scent of fecundity, a childhood memory of my dad's compost heap. Chad keeps on poking around in the hole with his fingers, looking for worms and signs of fungus. Biological soil, he tells me, is one of his two passions. His other is water.

'Climate change is fucking bullshit', he says in a soft drawl, not aggressively, more in resignation. 'The allocations that are given out are made to deal with climate change. They deal with climate change because the climate changes every year. One year we've got drought, the next we have flood. The allocations cater for that.'

The allocations Chad is referring to are the water entitlements of the St George irrigators. There are two sorts: those for harvesting floodwater and those for pumping water from Beardmore Dam and Jack Taylor Weir. Chad's entitlement is to harvest floodwater. These licences are only activated when the dams are overflowing. Then, and only then, Chad is permitted to pump water from the river. Cubbie Station, a cotton farm 80 kilometres downstream from St George, works the same way—Cubbie is notorious for being licensed to harvest and store more water than is held within Sydney Harbour. Years can pass between floods, but when the waters do rise, the irrigators can pump vast amounts of water into their on-farm storages. The other type of irrigation entitlement is far more reliable. But even so, if there is not enough water in the dam and weir, irrigators receive less than a full allocation. Critics say there are too many licences on the Balonne, and that the annual allocations and water harvesting entitlements are too generous. Chad disputes this.

He's a lanky, likeable American, with a keen sense of irony and an endearing openness. His still-dark hair and enthusiasm make him seem younger than his fifty-seven years. He tells me he came to St George as a pioneering 21-year-old back in 1972, the era when Queensland's dams and weirs were being constructed and the government of Joh Bjelke-Petersen was mad keen on developing the hinterland. Chad remembers St George as a frontier town. 'There wasn't even an automatic telephone exchange. I'd call the states and the operator would be listening in and interrupting all the time. "Three minutes. Are you extending?" Things have come a long way since then.'

He takes me to see one of his farm dams. It's breathtaking. A huge flat sheet of shallow water stretching off towards the horizon. He quotes figures in megalitres as I try to take it in. The dam is further across than Canberra's Lake Burley Griffin. Chad knows exactly what he's showing me and how photos of these dams have been splashed across the front pages of an indignant metropolitan press, turning Queensland's cotton farmers into latter-day pariahs. He wants to set the record straight, to get his point of view across. He tells me he filled this storage last summer for the first time in seven years, that he and other water harvesters are only able to take water when the river reaches flood levels. He's used half the water to grow a winter crop of irrigated wheat, constantly consolidating the remaining water into smaller and smaller storages to reduce evaporation. Now he has enough left to produce a summer cotton crop. He makes a very clear distinction between water harvesting, which removes massive amounts of water from the system but only at times of flood, and the more modest annual irrigation allocations that allow water to be pumped from the river. Some farmers have allocations, some are solely water harvesters, and some have access to both.

'The allocations, they're the foundation of this town. And when everything goes to shit here like it has done for the last seven years, and there's no money anywhere and no water anywhere and none of the flood harvesters grow a crop and the dry-land farmers have the seat out of their pants, irrigators out here might not have had a full crop every year, but they had a crop every year. And that's the basis that keeps all the businesses in town going. Those dry-land farmers that are hauling in the wheat this year, they have the tractor dealerships and the banks and all the rest because of those thirty or so properties that have allocations.'

But what about the flood years, like last summer when water harvesters could fill their storages but the Culgoa down in New South Wales didn't get enough water to flow out onto the flood plain? He says that's nothing to do with water harvesting but rather is determined by the duration of flooding. He says a flood with a high peak but a short duration will always dissipate before spreading out across the flood plain, while a flood with a lower peak but a longer duration will have the staying power, regardless of the water harvesters, to push out through the channels to the Darling.

We head back towards his office and I ask him what he'll do with his property—Balonne Plains—when he retires, whether he'll hand it down to his son. 'Not likely', he says wryly. 'He works in the environment department in Canberra.' So has it all been worth it: the shift from California, all the hard work, the effort, the opprobrium? His response is typical farmer philosophy. 'Oh yeah, it's been worth it. Some people are happy in Siberia and others are miserable in paradise. Shit happens wherever you are; it's just the flavour that's different.'

Sitting behind his desk in his farm office on the other side of town, Richard Lomman is not so philosophical. He's pissed off, irritated by the bad press that irrigators, particularly water harvesters, keep getting in the big city newspapers. Outside, 260 hectares of vines growing table grapes shine green and vibrant in the sun, watered by a mixture of the more reliable river allocations and a less reliable water harvesting licence. Richard doesn't own this property but he's managed it for eighteen years, building it up from scratch, putting in the large water storages that have caused so much angst. Unlike Chad Prescott, he doesn't have the luxury of choosing not to plant a crop if there's no water—he needs to keep his vines alive through good times and bad. He doesn't hold a high opinion of city dwellers who eat his grapes and wear his neighbour's cotton and then turn around and criticise the accomplishments of St George. He's proud of those 260 hectares of grapes and what they do for St George.

'We employ about eighteen permanent staff and up to 300 at harvest time', he says. 'Our wages bill is about $5 million per year. I don't know what the multiplication effect of that is, but a fair proportion of that would be spent in town buying food, buying clothes, pubs, restaurants, fuel.'

There is a widespread belief around St George that to let water run down to spill out on the flood plains of the Culgoa and the other delta rivers lying between the Condamine-Balonne and the Darling would simply be a waste. Lomman has little sympathy for the graziers south of the border and he's unapologetic about redefining the flow of the river. And he has a powerful ally. It's the Queensland Government, which ultimately decides just how much water can be taken from Beardmore Dam in any given year, and sets the rules governing the activation of flood plain harvesting.

'We're changing the landscape because we're here, and we're going to continue to change it. We can't expect it to be the way it was 100 years ago. Those guys down there [in New South Wales] have to realise that there's a huge economic thing going on. There are communities up here that have been here probably nearly as long who are now reliant on the water we have. So if we take all that water out of our community and send it down there, all we are doing is transferring security from this community to that community. This community dies.' And Richard Lomman crosses his arms across his chest, challenging me to argue otherwise.

⌒

The next day I drive out of town, heading west; I've got a long way to go. I think about St George, 'Inland Queensland's Fishing Capital'— about its sprinklers, its plaques and its pioneer mentality. I'm not really in a position to pass judgement. After all, I live near the shores of Lake Burley Griffin, an artificial lake also held back by a concrete dam; a lake three times the size of the lagoon backed up behind Jack Taylor Weir (although a third the capacity of Beardmore Dam). Burley

Griffin is an ornamental lake, set there to impress visitors, give the bureaucrats something to jog around, and soften the concrete angles of our national institutions. It provides few jobs and underpins no regional economy. And yet, much as I should admire the irrigators and what they have achieved, the image that returns to my mind again and again as I drive away from the Condamine-Balonne-Culgoa and the Darling's upside-down delta, is the look of bewilderment etched into the face of a young National Parks ranger as he searches in vain for a waterhole in the waterless bed of a desolate river.

2

BOURKE

The Darling, south of Bourke

It's late in the day, my work is done, but instead of relaxing I'm tearing along a dirt road, pushing the four-wheel drive as hard as I dare. Emus scatter across the red-soil plain and into the scrub, fleeing the roaring car and its kilometre-long dust plume. 'Can't be much further.' I repeat the mantra: 'Can't be much further'. For off on the western horizon, a wall of storm clouds is rolling rapidly east through the sunset to meet me, trailing long curtains of translucent rain. If it catches me out here on the dirt, I could be marooned for days, four-wheel drive or no four-wheel drive. As if to emphasise my vulnerability, the road changes back to black soil, the powder-fine silt of centuries. It's as fine as talcum now, as light as flour, but once the rain hits it will instantly become a treacherous skidpan. A little more rain, the sort of serious rain the clouds are hauling towards me, and the road will turn into sucking, cloying mud, more than capable of bogging bigger vehicles than mine and better drivers than me. I'm on the wrong road at the wrong time, lulled into complacency by days on end of unblemished skies. Now there's little choice; I inch the speed up another 5 kilometres per hour, willing my hands to stay light on the steering wheel even as the car drifts from side to side across the floating surface of the dead-straight road. An unwelcome memory of a car wreck lying on a claypan somewhere to the south flashes through my mind. Off to one side the scrub begins to move, then the storm winds engulf me as well. My ears pop as the air pressure drops. It won't be much longer now. A raindrop splatters into the windshield, leaving a circle the size of a saucer. 'Can't be much further, can't be much further.' And then I see it in the distance, the tiny triangle of salvation: a give-way sign. Not a moment too soon. I ease off the accelerator, tap the brakes and drop back a gear, slowing gradually before rolling onto the bitumen of the highway with a sigh of relief. And in that moment, the storm front is transformed from a thing of menace to the embodiment of beauty.

Back in Canberra, down in Australia's south-eastern corner, the rain clouds come in low and dark: the lower the clouds, the more chance of rain. They come up from the Great Australian Bight, bringing the same rain that falls on Melbourne or Sydney, or they sweep inland from the Pacific in localised fronts. Either way, when it falls, the rain closes in around you, blocking out the world. But out here, 50 kilometres east of Bourke, this rain is something new. These clouds, arcing lightning, are high—kilometres high, impossibly high. Moving en masse across the sky, they're the tail end of the Asian monsoon, moving down from Indonesia, over the Timor Sea and out across Australia's central deserts. If it rains in Alice Springs, two days later it rains in Bourke. The storm front has taken no-one by surprise, no-one except me.

Secure on the bitumen, I stop the car and climb out, alone on the empty plain. The air crackles with anticipation and the first drops smack into the road. The clouds are so high that the setting sun is shining beneath them, backlighting the translucent curtains of rain, making them shimmer gold below the grey as they unfurl above the dun-coloured earth. It's an aurora of water and sunlight. The curtains hang diaphanous, opening up the world instead of closing it down. Some reach the ground, some fall short; others retract slowly back into the heavens even as others unwind earthwards. Time changes: the individual curtains move with majestic slowness, yet every second brings some new revelation. Lightning bolts dance and there's the sound of thunder, strangely distant. A curtain is coming my way. I can see it moving up the road towards me. Now I can hear it, the advancing water hitting the pavement with a roar. Another gust of wind and it's upon me, the huge drops stinging as they hit. I dance around on the road, arms out, exhilarated. The water buckets into me, soaking me through. Then the curtain has passed and the road steams in appreciation. The sunlight catches the steam, a small facsimile of the glowing auroras of rain.

By the time I get to Bourke the sun has passed behind the clouds and the rain has become grey and mundane, if rain in the Western District can ever be called mundane. I head into the main street and the residual exhilaration drains away. I've seen streets like this before: in the Middle East, in South America, in Asia. The main street of Bourke is locked down for the night. There's hardly a windowpane left exposed. Entire stores lie encased behind roll-down steel shutters, security doors, padlocks and chains. It means the same thing in every place I've seen it: poverty and crime. Bourke, like Walgett and Brewarrina, has long been a home to Aboriginals driven from the land or liberated from the missions; many succumb to a cycle of grog, gambling and welfare dependency, along with the domestic violence and petty crime that goes with it. In Bourke's main street, only the police station, its lights on and four-wheel drives out front, is unshuttered and open for business.

The rain pauses. Perhaps the front has passed. I should find somewhere to stay, but first I decide to pay my respects to the river, running parallel to the main street a block over. My first sight of the Darling cheers me up again. It lies deep and broad down in its channel, the river gums strong and healthy, the way a river should look. Although the town is just metres away, the banks are bush-lined and surprisingly free of litter. It's a river from another time. I know it's an illusion, that this long stretch of water winding through the eucalypts is held there by the town's weir, somewhere out of sight to the south. But after the empty Culgoa and the bloated weir-pond of St George, it's comforting to see something that at least approximates the dictionary definition of a river. I walk to the town wharf, a multistorey wooden structure rising many metres above the river. It's a reminder of just how high the waters once ran, back in the days when paddle steamers would haul the wool clip all those hundreds of kilometres down to the Murray. It's not the original wharf, of course. Built in 1996, it's just the latest replica, the fourth, of that which gave Bourke its purpose. It's there to service tourists, not graziers.

Bourke is no accident. It's located where the rivers and channels of the Northern Basin finally consolidate themselves: the Darling river begins just east of town, where the Condamine-Balonne-Culgoa enters the Barwon, which itself has already subsumed the rivers flowing from the east and south, including the Macintyre, the Gwydir, the Namoi, the Castlereagh, the Macquarie and the Bogan.

In the car park by the wharf I encounter two grey nomads, chatting amiably of their travels beside their huge motor homes. The two women, together with their husbands, have been plying a similar route, part of the elderly tide sweeping south from Queensland to outrun the heat of the coming summer. This is the third time they've encountered each other in the past week.

'Bourke? Yes, we like Bourke. Usually call in here for a day or two', one lively sexagenarian tells me.

'Really? Main street doesn't look too welcoming', I suggest.

'No. Not when it's all shut up. Just wish they had some public toilets.'

Back on the main street, R.W. & E. Culhane, Clothing and Footwear has a sign up: 'Toilets—Tourists Only', a last echo of a previous era. I climb back into the car and head off to find somewhere to stay.

⁓

The next day I'm stuck in Bourke. I wasn't planning to stay, intending instead to head into the outback west of town, out to the Paroo. But I'm not going anywhere, not yet. The bitumen ends at Bourke, and the dirt roads running west of town, to the 'back o' Bourke', have been rendered impassable by the rain. The police have closed them and nothing is allowed through: not road trains, not four-wheel drives, and certainly not the likes of me. The rain has left its legacy, puddles of water lying beside the road, but the day has dawned clear. The early morning sun has a fierceness to it, a determination to make up for the previous afternoon's lapse. In another day or two the roads will be dry enough for me to continue on my way. In the meantime, Bourke beckons, the iconic town of the Australian bush.

Business hours have transformed the main street. The shutters are gone, the padlocks have been put away, and cars and trucks line both sides of the street. Only Fitzgerald's Post Office Hotel, boarded up for good, gives the lie to apparent prosperity. I drop into the local Retravision shop to pick up a cheap radio. It's a modern electrical goods store, indistinguishable from those found in any suburban mall: wide-screen TVs, digital cameras, iPods. Well, almost indistinguishable. On the counter, next to the photo-printing centre, is aligned a neat display of stubby holders made from the same rubberised material used to make wetsuits. It's a promotion: customers can have their digital photos printed on their own customised stubby holders. The ones on display feature smiling teenage girls, their hair brushed and their teeth gleaming white, like extras from an American sitcom. Their teeth are the brightest part of the pictures, their teeth and the iridescent red guts of the eviscerated wild pigs they're posing with. Other customers come and go. No-one looks twice at the stubby holders.

Bourke's boom years were back in the second half of the nine-teenth century, when the Australian colonies were riding out of obscurity and into prosperity on the sheep's back. Vast grazing pro-perties in the Western Division, 1000 kilometres from the sea, were generating unprecedented wealth. But they needed to get the wool clip to market, and Bourke provided the means. A port was established after the first paddle steamer reached the town in 1859. Afghan camel handlers competed fiercely with oath-spewing bullockies to bring the wool in overland to Bourke, where it was loaded onto the steamers and floated off down the Darling on a wing and a prayer. It was a haphazard affair; then, as now, the river could go dry for months, even years at a time. Drought would strike and the steamers would run aground in the channel, stranded for small eternities until the next flood refloated them and sent them bobbing down the river. There are the remains of a boat downstream from Bourke that chased the floodwaters 22 kilometres out across the plains to collect its cargo, only to be left stranded there for all time as the waters receded. Or so the story goes. You hear it everywhere along the Darling and down the Murray: same story, but the location of the boat is always somewhere else, somewhere not too far away, but not so close either. A riverboat, *The Jandra*, still plies the Darling at Bourke, but it's restricted to the few kilometres of the river held above the weir. It, too, is a replica, built in recent times to carry tourists, not cargo.

The riverboat era all but ended in 1885 when the railway line reached town. Bourke was no longer just another town on the Darling; it was *the* town on the Darling, its status underwritten by the railhead. For the railway didn't just cart wool off to distant ports. It brought services the people craved: regular post, the latest fashions and newspapers; dry goods, theatrical troupes and the occasional politician. Bourke was no longer an outpost but a garrison town, a fortress of civilisation beyond which lay only the great untamed land. And a new name emerged as shorthand for that land: the back o' Bourke. For on the train came newspapermen and writers, eager to feed a growing demand for frontier news. The name Bourke recurred time and again as readers round Australia followed the tales of colourful swagmen, sly-grog merchants and ne'er-do-well remittance men; of good girls down on their luck, fearsome natives and gallant bushrangers—all living larger than life in the rugged frontier back o' Bourke. In the world of commerce, Bourke was indeed the centre of the Western Division; but in the world of the imagination, it was emerging as the capital of a mythological kingdom known simply as 'the bush'.

The railway ensured Bourke's prosperity well into the twentieth century, but by the 1980s wool was in serious decline. Then, in 1990, the railway era ended even more abruptly than that of the paddle steamers. Ironically enough, the cause was a flood. Bourke shrugged off the rising waters but they wreaked havoc on Nyngan, 200 kilometres to the south-east. The flood backed up behind the railway line, inundating the town. Almost the entire population was airlifted to safety by helicopter. In a last-ditch attempt to lessen the impact, the Army dynamited the railway line north of the town. And that was that. Rebuilding the line was deemed uneconomic, and Bourke lost the thing that had set it apart. Road trains replaced the locomotives overnight, simply passing the town by as they thundered off to the coast.

And so Bourke returned to the river, and to irrigation. Elsewhere, in Queensland and further east, good money was being made from cotton, and Bourke had already followed the trend, drawing on the Darling to grow the crop, as well as grapes and citrus. Throughout the 1970s and 1980s the state government handed out water licences, encouraging irrigators to become bigger, more diversified, more water intensive. And for a few short years it worked and Bourke boomed again. But it was never going to be sustainable. For the water coming down the river—and it had never been that much or that regular—was dwindling year by year as farmers on the Condamine-Balonne, on the Namoi, on the Gwydir and on half a dozen other tributaries syphoned it off before it could reach Bourke and the Darling. By the early summer of 1991, little more than eighteen months after the flood that submerged Nyngan, the flow in the Darling had become so stagnant, so enriched by agricultural chemicals, that a toxic blue-green algal bloom stretched uninterrupted for 1000 kilometres. Nothing could touch the water: not stock or humans; not to drink or swim. The ABC helicopter swept low over the fetid river, sending back devastating images to disrupt the Christmas-time complacency of faraway capitals.

The river was dying, poisoned, depleted. Committees were formed, research conducted, reports written. And eventually action was taken. Caps were placed on how much water could be taken from the river. But rather than save Bourke, the new caps punished it, for nowhere were they more onerous than on the upper Darling.

And still Bourke managed a fair imitation of prosperity, the irrigators making do as best they could. But in 2002 drought began, as bad or worse than the Federation drought that had left the town gasping a century earlier. The cotton growers stopped being farmers and businessmen and became gamblers, staking what little water they had left to plant a crop, punting that summer floods would bring more down the river in time to bring the plants to maturity. But the floods stayed away and the gamblers lost their bets, then their shirts and finally their farms. Back o' Bourke Fruits, one of the town's biggest employers, struggled to get enough water to keep its grapevines and orange trees alive. In desperation, the town council sacrificed some of the townspeople's own water, the water meant for drinking and washing and cooking, and gave it to the company in a last-ditch attempt to keep the plants alive. But it was too late. The drought went on and on and on and the oranges withered on the trees. Back o' Bourke Fruits went belly up. Workers, skilled and unskilled, started leaving town for the mines, while those lacking the skills or the motivation went onto welfare. And increasingly, Bourke went back onto the government teat: Centrelink, the Australian Tax Office, Medicare, Aboriginal legal and health services, the catchment management authority, the National Parks and Wildlife Service.

⌒

I head out of town to one of the big irrigators, Darling Farms. On the way I pass what's left of Back o' Bourke Fruits. On the back road, piles of dead vines and bulldozed trees lie black under the sun. Out front some vines remain alive, defiantly green, a brave facade maintained by the administrators to lure prospective buyers. Twenty kilometres from town I turn onto a red-soil road that takes me through the mulga and past Darling Farms' huge machinery shed, a jet ski sitting incongruously outside, to the property's office. It's in a building the size of a large house. Inside I find Steve Buster, working through the figures, the figures that steadfastly refuse to add up. The office complex feels empty, and not just because it's a Saturday morning. The property used to employ five managers, two agronomists and an accountant. Now Steve and his brother-in-law Ian divide the work between them, running the farm for the bank while it seeks a buyer. Steve takes me through the figures: the investments, the debt, the water allocations. The numbers aren't for publication but he's not disguising what has happened here: it's not a pretty story.

Steve Buster is forty-five. His face is farmer pink, his hands calloused, but he talks with the modulated voice of the university educated. And through that voice intermittently drifts the soft inflections of southern California. In the early 1960s Steve's parents left the United States and settled in Bourke, where they bought a clapped-out grazing property on the banks of the Darling and started growing irrigated cotton. Steve tells me how enthusiastically the state government welcomed the immigrants, how keen it was for irrigation to flourish at Bourke, how water licences were there for the taking, right through the 1980s and into the 1990s.

'We were encouraged to develop and diversify the farm and utilise the resources we'd been given to the point that the threat was that if you did not develop your water licence it would be taken from you. So this farm continued to develop right through the late eighties. In 1991 we were fully developed, and we grew about 3000 hectares of diversified cotton', Steve recounts.

But not until we jump into his truck do I begin to understand the scale of Darling Farms, and the scale of what has befallen it. We drive to Gidgee Lake, not a natural lake but a huge artificial water storage, held above the surrounding fields by massive levees. The lake stretches off into the distance, 1.5 kilometres wide, the far shore a thin dark line separating water from sky. Dead trees stretch their limbs clear of the water, remnants of the flood plain from the time before the lake's walls were bulldozed from the plain. The trees are dead, but not the birds: pelicans and gulls wheel away, at home in the desert heat. A steel walkway protrudes out over the lake's surface; a measure indicates the water is 5 metres deep. But Steve tells me Gidgee Lake was empty for two years before floods last summer gave the Busters the opportunity to fill their storages. Half the water has already been used on a winter wheat crop. The rest, consolidated into Gidgee Lake, will go towards this summer's cotton crop. By mid-year the lake will be empty again unless another flood comes to fill it. We track around the edge of the lake as Steve explains how the water is reticulated, how the already flat land has been laser-levelled. We come to a boat ramp where a couple of catamarans lie in the sun, awaiting their crews. A decade or so back, when the storage was empty, one of his managers saw the opportunity to clear a swathe of dead trees from the bottom of the storage to build a sailing course. Steve looks out over the lake fondly and tells me how, on a hot day after school, his 15-year-old son takes a powerboat and goes water skiing with his mates. 'A treasured life,' he says, more to himself than to me, 'a treasured life'.

Moving away from the lake we drive alongside massive irrigation channels, channels so large that back in the days when there was enough water to fill them, if the farm kids grew bored with water skiing on the lake, they used to tie ropes to the backs of trucks and ski the canals instead. But now the canals run half-empty. Too much water is lost to seepage and evaporation, but there's no money left to re-engineer the farm with smaller, more efficient canals. We drive to the river where eleven massive inlet pumps patiently sit, each dipping a 60-centimetre pipe into the water, each capable of moving 100 Olympic-size swimming pools a day. Darling Farms is a floodwater harvester; its owners can only take water when the river reaches a certain height. So when it gets there, they pump for all they're worth.

Steve takes me past a wheatfield where a header is taking off the irrigated crop. It's been a good result but the money goes to the bank now, not the family. And then he shows me something I haven't asked to see, something I never knew was there. He takes me past rows and rows of jojoba, sickly and parched in the sun, and through fields of dying citrus, with an orange or two hanging forlornly amid browning leaves. Each tree is drip-watered, electronically monitored by state-of-the-art probes, but the water was turned off months ago.

'We were encouraged to diversify', says Steve. 'We had 130 hectares of citrus. A full-time work force of fifty, another sixty-five or so casuals. Now look at it.'

'How much did it cost to put in?' I ask, looking non-plussed at the dying trees.

'Around two million', Steve says matter-of-factly. He puts the truck into gear and we move on.

We pull up next to what looks like a large greenhouse, except instead of glass, the structure is covered in fine netting to keep the birds out. Inside he shows me other experiments, now dying off like the orange trees and jojoba—asparagus, mangoes, other plants I don't recognise: all abandoned. 'We kept it all going for a while, but we were told prospective buyers didn't want to know. They're only interested in cotton and wheat and crops you can grow if and when the water is there. No-one wants permanent plantings.' He takes a last look around and we return to the truck.

Steve Buster is middle-aged. Darling Farms has been his life, the life of his parents. To some, cotton farmers like him are the enemies of the river, sucking the top off floods. And yet I can't help but admire the man. Drought, government policy and, no doubt, his own poor judgement have robbed him of his livelihood, his parents of their home and his children of their inheritance. Over almost half a century, the Busters have sunk tens of millions of dollars into Darling Farms, and they will walk away with little or nothing to show for it. The bank will take it all. And yet here is a man willing to take me on a tour of his failures, of his shattered dreams, of a birthright wrenched from his grasp. I put myself in his place: would I volunteer to be a tour guide to my own downfall?

'I don't hold onto things in this earth too tightly', he says. 'I'm grateful I wasn't born in Bangladesh. I'm disappointed for my kids' sake. It's a treasured life, but God is no man's debtor. The measure of your life is not how many toys you have. So you want to keep a bigger perspective in this.' There is a prospective buyer in the wings. Steve believes that, before another year is out, he will have left the farm. Already he's putting out feelers, looking for employment managing someone else's dreams, or consulting on the harsh realities of modern agriculture.

We head away from the irrigated fields, off through the mulga scrub, bashing the bush down as we go. There's one more thing Steve wants to show me. We pull up among the car-high scrub, the exposed soil glowing Martian red between the sporadic vegetation. In front of me is a long thin slab of concrete, its surface pockmarked by the years.

'Do you know what it is?' he asks.

'No idea. Building foundations?'

'Nup. Cricket pitch. They used to play here up until the sixties.'

'You're kidding.' Around me the head-high scrub is everywhere.

'Shows you how quick the bush takes back its own', Steve says.

I drive back into town. Someone has draped a dead hare over a school-crossing sign. Charming. I go past the Central Australian Hotel, closed and boarded up, overlooking where the railway once ran; past the SPAR supermarket—'World Class—Australian Owned'—with its steel shutters now pulled down for good; past Kazza's Cafe and Pizza, burnt out and abandoned, with a frayed police crime-scene ribbon fluttering in the breeze. Not so long ago there were seven pubs in Bourke, now there are three; where there were five supermarkets, now there is one. The town is down to its last Chinese restaurant, housed upstairs in its last remaining club, the bowls club. The RSL shut for the last time a few weeks ago, a padlock on the gate and a 'for sale' sign on the fence. Over in the park the lists of the war dead remain: four in the Boer War, thirty-three in World War I, nineteen in World War II. Country towns were the heart of the Australian Imperial Force, in spirit if not in numbers. Many hundreds of men went to war from Bourke. And now the RSL is gone.

I call into an employment agency, one of the few growth industries in town. Regional Manager Jackie Davis, a large friendly woman in her mid-forties, manages a staff of eight. Bourke born and bred, she welcomes me into her office. She took the job when the bottom fell out of the local taxi industry. She recounts how her father, John, ran three taxis back when wool was king. 'People would get in a cab and go to Melbourne. He did a couple of trips to Cairns. A shearer might have had a windfall, or a couple wanted to go down to the races. And now, they're struggling to put fuel in their car.' The town has one taxi now, and it doesn't operate after nine at night.

Jackie says it's become more difficult to put people into jobs since the decline in irrigation, especially unskilled workers, many of them Aboriginal. 'Back in 2000 and 2001 we were very active', she says. 'We had all the citrus and grapes. We were taking out busloads of people of a morning and back. We'd have them lined up wanting to register to go out to get seasonal work. It's what the community here liked, not the nine to five sort of work. We were a seasonal town.' I thank Jackie for her insights, and before I leave I check out some of the jobs on offer. There are a few in Aboriginal health, a few in aged care, one for a part-time cleaner. Someone is looking for an experienced stockman, preferably a couple: 'must have own transport and dogs'. The job is 50 kilometres from Goodooga and the pay is $15 an hour plus free meat (one sheep per fortnight). I do the sums: $30 000 a year for a couple, accommodation in the middle of nowhere, and twenty-six dead sheep.

I return to the main street, the one place that seems to remain undaunted by the drought. I wander into R. W. & E. Culhane, Clothing and Footwear. Spread over two normal shops, it's a clutter of clothing and homeware, a sprawl that's evolved over decades of country town retailing. In among his inventory, in a small office in the middle of the shop, I find Bob Culhane, former town mayor, former president of the Chamber of Commerce, another who's Bourke born and bred. Now sixty-seven, he's been running his store on the corner of Oxley and Sturt Streets for thirty-five years.

'Up to about five years ago we had the town bouncing along pretty well, but then the drought has just wiped us out', Bob says. 'Now we're trying to survive. The town has shrunk by about 25 per cent. The people who have gone are those that were earning their $1000 a week working on the fruit and the cotton. The breadwinners. So more than 25 per cent of the cash flow has gone; a good bit of what's left is on the dole. I'm at the age where I want to retire. One of the options I'm looking at is just walking out, closing the front doors and walking out. Our turnover would be down at least 50 per cent from where it was five years ago. I don't know if I could sell it. We've put ourselves on a pension, drawing from our own super fund, just to keep the doors open.'

Bob is matter-of-fact. No self-pity, no finger pointing. It's just the way it is. I get the impression this is going to be a short conversation. Until I stumble upon the right question. I ask whether there's any-thing special about Bourke, anything worth saving. Bob's eyes light up. 'Oh yes', he says. 'Bourke is a special town. When I was younger I used to bone kangaroos for a living. The roo works here closed down and I went over to Brewarrina to do the same thing. The Brewarrina blokes used to go down to Sydney and they'd get the shits, because people would say "Where do you come from?" and they'd say "Brewarrina" and people would say "Where's that?" and they'd say "It's close to Bourke". And people would say "Oh yeah, Bourke, we know where Bourke is".'

The story leads to another and then another. We leave the here and now and its depressing statistics behind, and in his quiet understated manner, Bob takes me back to his town's glory days, back to when the railway still operated and the town was the end of the line, to an era when, every now and then, a severely hung-over young man would stagger dumbfounded from the train, vaguely recalling a Sydney bucks night, adamant he must get back in time for the wedding.

'There was a bloke, Wobbly Jackson, a top horse breaker', says Bob, starting another yarn. 'Had a bit of a speech impediment, but he was good with the horses. He got this lovely horse and trained it up. He gave it to his wife Valerie to ride, and they went out riding one morning. But the horse bolted with her on it. Wobbly raced after her, shouting "Walerie, Walerie, wide it into the lake, wide it into the lake!" Anyway, she rode it into the lake, and that stopped it, and everything was okay. Then, next day, Wobbly takes the horse out for a ride himself and sure enough it bolts. Well, he thinks, bugger this, so he pulls out his gun and shoots it dead, right in the back of the head. Just like that.'

'What, at full gallop?' I say incredulously. 'Lucky he didn't break his neck.'

Bob gives me a disappointed look, and then says, 'They breed 'em tough out here'.

And in the pause that follows, I realise my mistake. This is a bush yarn, not a rendering of history. That's the essence of them: they stretch the truth, the more outlandishly the better. Out in the relentless heat and soul-sapping monotony of the nineteenth century, where one day merged into another, the bush yarn relieved the boredom. The old stockmen, sitting around their campfires, recounted the same yarns again and again, each time with some fresh embellishment, a conspiracy of embroidery. Who would question the story of a paddle steamer marooned halfway to Alice Springs? Only a pedant. I put my city-bred scepticism to one side and listen.

'In the old days,' Bob continues, my transgression forgiven, 'the swaggies used to walk up and down the river. They'd do a bit of shearing, a shed here, a shed there, and they'd grab their money and keep going. And if they started getting a bit skint, what they used to do was cut some wood and leave it in a pile on the riverbank for the paddle steamers. And they'd leave a little tin there. And a captain would come along, use the wood to fire the boilers, and put the appropriate amount of money in the tin and take it along to the next collection point, a pub or an outstation or something. And the next time the bloke was passing he'd go and pick up his tin with his money in it. Anyway, one time down on the Murray, one of the captains took the wood but didn't leave any money. So next time this captain came through, the same bloke who had cut the wood the first time saw him coming, and this time he left a stick of dynamite hidden in the wood. And it went into the boiler and blew the boat to smithereens'. And this time I know better than to make uninvited observations.

Back when Australia was young, back in the late nineteenth century when it was still a grab bag of colonies and Federation a dream, the bush yarn helped shape a nascent national identity. Australia had fought no wars, so there was no well of ANZAC spirit to draw upon; daylight bathing was outlawed, so the egalitarianism of the beach was yet to emerge; and national cricket teams were only starting to visit England, so the notion of sporting prowess was still in its infancy.

Unlike the Americans, with their War of Independence, constitutional rights and finely wrought words of liberty and freedom, Australians had precious little to build a nation upon. So it fell to the bush, where Jack was as good as his master and mateship was the common creed, to provide the kernel of identity for a new country. Down in the coastal capitals, the colonial establishment looked to Britain for life's template, importing hedgerows and foxes, rabbits and oak trees, attempting to manufacture a new England. Less so the working classes, where republicanism stirred the larrikins, the Irish settlers and the descendants of convicts. They turned to the bush for inspiration, creating a romanticised vision of the never-never where men were noble, unbent and free. Bushrangers, violent and ruthless, became folk heroes while their city counterparts remained simply criminals; semi-literate swagmen, footsore and hungry, embodied a freedom that could never be found in a city factory; striking shearers, often small landholders themselves, symbolised resistance and solidarity at a time when city unions failed to break the bull system on the docks, where each morning foremen picked out wharfies at their discretion for a day's work at a time, take it or leave it. Beneath the mythology, the importance of rural Australia was no fabrication: the wealth of the colonies flowed from the goldmines and the shearing sheds beyond the great divide, and seminal political events were unfolding in the hinterland—the great shearers strike of 1891; the formation that same year of the Australian Labor Party under a gum tree in Barcaldine; the 1893 Federation conference in Corowa on the Murray. It seemed that the bush, with all its bullshit and bluster, might just provide the impetus for a new and better land.

In 1880, JF Archibald established the *Bulletin* magazine in Sydney. It was nothing special, just the usual dreary bumph about politics and business, until Archibald began to devote pages to readers' contributions: their poetry, yarns, reports of everyday life. In no time, bush yarns came to dominate the magazine; it gained the epithet 'the bushman's bible' and ran the motto 'Australia for the Australians' under its masthead. The *Bulletin* championed the good (egalitarianism and mateship), the bad (protectionism) and the downright ugly (racism). But above all else it mythologised the bush, finding there all the qualities and virtues to which a young nation might aspire. The *Bulletin's* own writers, most notably AB 'Banjo' Paterson and Henry Lawson, led the way. Paterson wrote 'Waltzing Matilda' in the wake of the shearers' strike, the ballad of a down-on-his-luck swaggie who steals a sheep, then commits suicide rather than surrender to a squatter and his tame troopers. It became the unofficial national anthem, confounding generations of foreigners. Lawson was less subtle, writing 'Freedom on the Wallaby' for Brisbane's *Worker*:

> So we must fly a rebel flag,
> As others did before us,
> And we must sing a rebel song
> And join a rebel chorus.
> We'll make the tyrants feel the sting
> O' those that they would throttle;
> They needn't say the blame is ours
> If blood should stain the wattle!

It's not this stirring last stanza that appeals to me, but rather the inspired doggerel of the first verse:

> Australia's a big country
> An' Freedom's humping bluey
> An' Freedom's on the wallaby
> O don't you hear 'er cooey?
> She's just begun to boomerang,
> She'll knock the tyrants silly,
> She's going to light another fire
> And boil another billy.

It would make Austentashus proud.

Modern-day historians, including the conservative Geoffrey Blainey and the left-leaning Manning Clark, credit Paterson and Lawson with helping to shape an Australian legend based on the ethos of the bush. Clark, in his *A History of Australia*, wrote:

The *Bulletin* encouraged Australians to believe in themselves ...

The *Bulletin* taught Australians to like the way they talked, to like the way they walked, to like everything about their country, to believe that Australian English, both in its spoken and written form, was a magnificent medium for the communication of what it was like to be a human being in Australia.

Clark was particularly taken with Lawson. He wrote a book about him, *Henry Lawson: the Man and the Legend*, and there is page after page on the bush poet in Clark's history:

> In the mighty bush [Lawson] had absorbed the wisdom of the bush people. He had learned that the Australian bushman communicated a vision of life by spinning a yarn. In short stories he began to tell Australians what they were and what life was like … He clothed the bush barbarians in a mantle of tragic grandeur: he had discovered the majesty and wonder and glory in places from which others for whom life was 'a tremulous stay' had recoiled in disgust, horror or contempt.

And if Henry Lawson helped shape Australia, then Bourke shaped Henry Lawson. In 1892 Archibald gave Lawson a rail ticket to the town and £5 for expenses. Lawson stayed out west for less than a year, but there he unearthed a rich vein, a vein he mined for the rest of his life. In Bourke, Lawson found a countryside gripped by drought, a far cry from the Arcadia so often depicted in the pages of the *Bulletin*. 'Blazing drought overhead and all around, burning the Darling banks to ashes, and audibly baking the land for a depth of several feet', he wrote, deploying the casual exaggeration of the bush yarn. The Darling was 'a muddy gutter', the shearing shed 'the most degrading hell on the face of this earth', and outback bushmen 'narrow minded, densely ignorant, invulnerably thick-headed'. 'We wish to Heaven that Australian writers would leave off trying to make a paradise out of the Out Back Hell; if only out of consideration for the poor, hopeless, half-starved wretches who carry swags through it and look in vain for work', he wrote in the *Bulletin*. And yet, as much as he apparently hated the country, Lawson found within the 'half-starved wretches' such a depth of humanity, with all its flaws, that he couldn't help but celebrate it. And in the hyperbole of the bush yarn Lawson found the true spirit of a people, a people prepared to laugh in the teeth of adversity, to find humour in the absurdities of life. Lawson, himself fated to alcoholism, deployed the excesses of the bush yarn in explaining Bourke's reputation as a hard-drinking town:

> When the Brewarrina people observe a more than ordinary number of bottles floating down the river, they guess that Walgett is on the spree; when the Louth chaps see an unbroken procession of dead marines for three or four days they know that Bourke's drunk. The poor, God-abandoned 'whaler' sits in his hungry camp at sunset and watches the empty symbols of Hope go by, and feels more God-forsaken than ever—and thirstier, if possible—and gets a great, wide, thirsty, quaking, empty longing to be up where those bottles came from. If the townspeople knew how much misery they caused by their thoughtlessness they would drown their dead marines, or bury them, but on no account allow them to go drifting down the river, and stirring up hells in the bosoms of less fortunate fellow creatures.

Lawson recounts how an entrepreneur from Adelaide collects Bourke's empties and piles them by the water to be shipped downriver. But he himself gets drunk, the river rises, and the bottles float away:

> They strung out and headed for the Antarctic Ocean, with a big old wicker-worked demijohn in the lead.
>
> For the first week the down-river men took no notice; but after the bottles had been drifting past with scarcely a break for a fortnight or so, they began to get interested. Several 'whalers' watched the procession until they got the jim-jams by force of imagination, and when their bodies began to float down with the bottles, the down-river people got anxious.
>
> At last the Mayor of Wilcannia wired Bourke to know whether Dibbs or Parkes was dead, or democracy triumphant,

or if not, wherefore the jubilation? Many telegrams of a like nature were received during that week, and the true explanation was sent in reply to each. But it wasn't believed, and to this day Bourke has the name of being the most drunken town on the river.

A while back I joined the *Bulletin*, then in its 127th year, to cover politics from Canberra. The editor flew me up to Sydney for some computer training and lunch. The computer consultant demonstrated how my stories would end up stored in a digital folder with my name on it. There on the screen were the folders of modern-day writers who, like Lawson, knew their way around a yarn: Paul Daley, Tony Wright, Julie-Ann Davies, Paul Toohey, Roy Eccleston and the rest. Intrigued, I clicked away. And there they were, folders marked 'Henry Lawson' and 'AB Paterson', their copy digitised for posterity, as if it had been written yesterday. A year later the *Bulletin* folded, a shell of its former self, killed off by the internet and apathy, left behind by a country rushing into a globalised future. And out on the western plains, Bourke, too, feels its mortality.

⌒

Before heading out of town I call in at the Port of Bourke, a large modern pub that wouldn't look out of place on Sydney's north shore or in Melbourne's eastern suburbs. It's owned by Clyde Agriculture, a UK-based agrobusiness and the last of Bourke's big irrigators. Inside, large black and white prints adorn the walls, depicting the halcyon days of the riverboat trade. It's late morning, not my usual hour for a drink, but Jackie at the job centre reckons it's when one or two of the old hands might drop in for a heart starter. And sure enough, I find old Mick Quarman alone at the bar, sitting on a middy of Fourex. I've never seen anyone able to sit on a beer like Mick. After chatting with him for more than half an hour, he's taken less than an inch off the top. He raises it to his mouth often enough, but apparently only to wet his lips. If it gets any further down his gullet, I see no evidence of it. Try as I might, I simply can't drink that slowly, and eventually my glass is empty. I order another and offer one to Mick. He up-ends his near-full glass and drains it in one mighty gulp, then places the empty on the bar. 'Thanks mate. Don't mind if I do.'

Mick's a former shearer, rouseabout and horse breaker, retired now. His skin is cracked leather, his blue eyes bleary with the years, his straggly grey hair hanging in threads from under his tattered Fourex cap. The old bushman tells me of his love for animals, of how he once worked a shed where the gun shearer could get through 250 sheep a day. 'He was a good hand, but he was angry. Always angry. The sheep could feel his anger. They'd fight him, and he'd fight 'em back. He'd shear 'em, but they'd be covered in cuts. By the end of the day he'd be rooted. I'd been breaking horses half my life, so I taught him to be gentle, to leave his anger outside the shed. After that, he was doing 300 a day.'

I ask Mick about the river, whether it's changed much. 'The river? Not what it was,' he says, 'not what it was'. And he tells me a yarn.

'About ten years back, I went down to the river. There was a bit of water in it in those days, and I put some nets in. Just to get a feed, you understand. Anyway, it was a hot morning and getting hotter, so I left the nets in and came in to do a couple of things. One thing led to another and it was a couple of hours before I got back to the river. Well, I couldn't believe my eyes. The net was chock-a-block, practically leaping out of the water. But just as I started pulling it in, out behind a tree comes this fishing inspector. He'd been hiding there, waiting for me to come back. "G'day Mick", he says. "Just getting a feed?" Well, I was gone for all money, and he stood there while I hauled in the net. There must have been 250 fish in there. But then it was my turn to start laughing, 'cos there were only three yellow-bellies and the rest were carp, and there's no bag limit on carp. Well, you should have seen his face. Anyway, I grabbed the yellow-bellies and went to head off. "Hang on a moment, Mick", he says. "What are you going to do with the carp?" And he had me there, 'cos it's against the law to put carp back in the river. I suppose I could've left 'em there on the bank, but it's a lovely spot and I didn't want to stink it out with a pile of rotting fish. So I loaded 'em into the boot, while this bloke stood there with his big ugly grin. But what he didn't know was that I'd thought of something to do with 'em, 'cos back home I had a few pigs you see, and I thought, "Well, they eat everything else, so why not a few carp?" And sure enough, I tossed 'em a couple and they gobbled 'em straight down. So I put the rest in the freezer and every few days the pigs would get a feed of fish. So I had the last laugh. Or so I thought.' And Mick pauses, lifts his glass to his mouth and wets his lips once more, knowing he has me hooked as well as any yellow-belly.

'What happened?' I ask.

'Well, at the end of the year, months after the last of the carp had gone, it was Christmas, and I picked out this one little pig and knocked it on the head. You should have seen it; it was a real beauty, fat on it like this, an inch thick. Anyway, I cooked it up for dinner and had a whole swag over. The pig roasted up something terrific, looked great. But when I served it up, I could see this look of horror come across people's faces. "Hey Mick," they said, "this pig tastes like fish!"'

By this time, Mick has skimmed mere millimetres from the top of his beer. But then he checks the time and drains the rest with a long steady gulp. 'Thanks for the drink mate', he says. 'I gotta go. Usually only have the one.' And he gives me a nod and heads out the door into the glowing heat.

A week or so has passed and I'm heading back through Bourke, heading home from Queensland. I'd been thinking of staying over in the old town, but it's still early in the day and there's a long way to go. So instead I drive straight through, pretending I'm just another passing traveller. The river, shored up behind the weir, looks wide, brown and benign. Someone has taken the dead hare down from the school-crossing sign and Bourke passes by unremarkably, just another country town. The road out of town is shadowed by the disused railway line; it passes the derelict abattoir that once employed 600 people and heads across the same flat grey plain that Henry Lawson described from his train carriage. I drive south-east along the dead-straight Mitchell Highway towards Nyngan.

I reckon Bourke will survive. It's got all those government agencies and, if nothing else, it can always trade on its history: the legacy of Lawson, the paddle steamers, and the sheep stations the size of small countries. And the bush yarns. A big new tourist facility, the Back 'o Bourke Centre, is almost complete, sitting above the flood plain and overlooking the river, with gleaming new toilets to entice the grey nomads. But it will never again be the town it was—the undeclared capital of the Western District, the heart of a mythical land that helped bestow upon the people of six far-flung colonies some notion of what it might mean to be Australian. That's all history now. In the week since I was last here, I've been out back o' Bourke, talking to the scattered souls who live there. More than one told me they no longer bothered with Bourke. One couple, from up on the Warrego, said that instead of travelling the two hours to Bourke, they prefer to drive six hours to Dubbo and make a weekend of it. Their kids get to stay in a motel, go to a movie, eat junk food at a shopping mall, and experience for a few short days what suburban kids take for granted.

I'm following the same road they must drive. It's a long flat journey, with nothing much to break the monotony: down through Nyngan, 'Floodtown 1990', with a commemorative helicopter planted outside the old railway station; down across the flat grey plain towards Dubbo. Lawson was right when he said there were no mountains in the outback. I'm almost at Dubbo when I spy my first hill in a fortnight. It doesn't boast much in the way of altitude, but at least it has some. And it's covered in trees, real trees, green ones, not bush scrub. It marks a boundary, an invisible line, where the smaller horizons of the east begin, and where dry-land farming again becomes viable.

And so, nearing day's end, I arrive in the new undeclared capital of western New South Wales. 'Welcome to the mighty tidy town of Dubbo', announces the sign on the highway. I drive in past the McDonald's, the KFC and the Pizza Hut, past the car yards and motels in their dozens, into the centre of town where the shopping trolleys of the Centro Mall career from Coles and Woolworths to Target, Flight Centre, OPSM and Gloria Jean's Coffee. History has been pushed aside: the old gaol has been dispatched to the end of a lane, its entrance crowded out by shops; a sandstone bank has been converted into a Hog's Breath Cafe, and the cars that drive past it are sedans, not farm vehicles. It's not so much a country town as the most far-flung of Sydney's far-flung western suburbs. There's a university campus, a

27-hole golf course, and an aquatic centre with a heated Olympic-size pool and waterslides. I pass the SunnyCove retirement home, where the bush comes to die, and the RSL, where the bush comes to marry. The RSL is massive, blanketing a city block, its banners declaring that Shannon Noll is appearing soon. The club has its own motel across the street and another building dedicated to organising functions, weddings mainly. There's a sushi bar in town and at least seven Chinese restaurants (including the Rising Sun Chinese Restaurant, complete with an emblem evoking the ensign of the Imperial Japanese Army—I'm guessing the proprietors aren't from Nanjing). On the outskirts along Wheelers Lane, out past Big W and Bunnings, new housing estates are eating into the wheatfields, McMansions on handkerchief blocks of land.

If Bourke is the past, then Dubbo is the future. I feel a touch resentful, *Bulletin* heritage and all, but it doesn't stop me revelling in my first espresso in two weeks or ordering a second to make up for lost time. After all, I live in Canberra, and what is Canberra if not Dubbo writ large? As my second coffee arrives, I pull out my maps, now creased and folded back on themselves, to plan the next day. And as I look at the lines and the dots, the names and the distances, I see something new. Civilisation is contracting eastward. The old garrison town of Bourke is now nothing more than an outpost, the railway gone and the river going. I see the new garrison towns, those with the base hospitals, the university campuses and the shopping malls, with the local newspapers, the multiplex cinemas and the air and rail links. They stretch in an arc down the eastern and southern side of the Murray-Darling Basin—Moree, Dubbo, Wagga Wagga, Griffith, Mildura—forming a Maginot Line beyond which the land and climate grow increasingly hostile. And as I consider the maps, I realise that every one of these towns is more or less dependent on irrigation for its economic base. But what happens if the water dries up, the way it has in Bourke? What then for these brave new suburbs of the bush? What happens if the drought really is the harbinger of climate change, if rainfall continues to decrease and dry-land farming is driven further and further east and south towards the mountains? Ross Garnaut, in his report to government, concluded that unmitigated climate change could halve irrigated agriculture in the Murray-Darling by 2050 and end it altogether by the turn of the century, causing widespread depopulation.

Dubbo, I realise, really is on the front line, and I contemplate it with a new respect. And I wonder if the worst does come to pass, and Dubbo, like Bourke, becomes a desiccated outpost, whether some future writer will mourn its passing, the way I find myself mourning Bourke. I hope so, but I doubt it. For Dubbo has not helped shape Australia, not the way Bourke once did: the mateship and the egalitarianism, the larrikinism and the gallows humour, the compassion and the resilience, the jingoism and the bullshit. I look for it, but nothing in Dubbo can rival the inexplicable charm of Bourke.

3

THE LAST WILD RIVER

The Paroo at Wanaaring in the early morning calm

Of the twenty-three river catchments whose waters inter-mingle to form the Murray-Darling system, the Paroo River is the healthiest. Survey after survey reveals that it has the best environment, the best river flow, the best fish life. Which is curious, given the Paroo receives the least rainfall of any of those twenty-three catchments. After all, if drought and climate change are sucking the Murray-Darling dry, then how is it that the Paroo is doing relatively well, out where rainfall is so slight and dry-land farming a distant rumour? The river's course runs the best part of 1000 kilometres from north to south, passing 200 kilometres west of Bourke through the outback proper, living and breathing on the edge of nowhere. It is the last wild river in the Murray-Darling. Wild, though, is a misnomer. For the floods of the Paroo are gentle, keenly anticipated and fondly remembered: life givers, not life takers. The Paroo is wild only in that it is unregulated, a bureaucratic word meaning it has no dams, canals or irrigation schemes, nothing apart from one minor weir along its entire length to disrupt the natural flow. The weir is at the Queensland village of Eulo, where a small date farm taps the main channel. The rest of the water runs unhindered from river beginning to river end. At least it has up until now.

The Paroo rises in the 'tin roof' country of western Queensland, north-west of Charleville, where the southerly extreme of the Asian monsoon falls on land so hard, so unforgiving, that instead of soaking into the earth it runs across the plain, funnelling into channels that in turn converge to form a watercourse that becomes definable enough to be given a name. When enough rain falls on the tin roof, a flood forms, surging south across the countryside. It pushes through Eulo and a week later it reaches Hungerford, 120 kilometres south, from where it crosses into New South Wales. It's a considerate, slow-moving flood, giving another week's notice before arriving at the next town, Wanaaring. No houses are inundated, no bridges swept away; the land is flat and the flood languid. Past Wanaaring, the land levels out even more, falling just over a centimetre for every kilometre of river. The so-called flood is now moving so slowly that it's possible to stroll along in front of it, waiting every minute or so for it to catch up, like some dawdling infant. Below Wanaaring it takes four weeks to proceed not much more than 100 kilometres. And still the land flattens, the silt from centuries of floods having levelled the plain, until the earth becomes so lacking in inclination that the Paroo itself loses what little purpose it has left, spreading out into a series of terminal lakes to die. But in death it's magnificent: the large shallow lakes explode with plant and animal life. After a good flood there are tens of thousands of birds comprising dozens of species, living and breeding miles from any permanent water. It's unique: a Murray-Darling river as healthy and life-giving at its end as at its beginning. Only once or twice in a lifetime will enough rain fall on the tin roof to give the Paroo the impetus to push beyond the lakes, 800 kilometres south of its watershed, and reach the Darling River. The Paroo is part of the Murray-Darling, but only just.

The dirt roads west of Bourke have dried enough for travel and I head out into the great expanse. The day is hot but not unbearably so, and after the rain the air is crystalline, the light hard and bright, with every detail, no matter how distant, held in sharp relief. The land is not flat at first, at least not if you're travelling east to west. Instead it rises and falls gradually, defined by long low ridges running north to south, as if the country carries a low wide swell washing in from the west. The weather has been kind this year, and intermittent rains have encouraged patches of grass where there's enough soil in among the mulga. Elsewhere the land is red and barren, impervious to growth no matter how much rain might fall. I stop at a claypan where spindly black tussocks stand incongruously in a couple of centimetres of collected rainwater. The water is going nowhere: not into the clay, not into any creek, and, by the look of them, not much into the tussocks. Another day or two and it will have evaporated. But for the moment, emus dodge skittishly round its edge, and small families of sharp-eyed goats graze by its shallows.

Rather than driving straight to Wanaaring, I detour along the Hungerford road to Fords Bridge, out on the Warrego, the first of the 'outside' rivers that run west of the Darling. It's a Sunday morning, I'm in no particular hurry, and some locals in Bourke have told me that Fords Bridge is a fine place to lose oneself of a weekend: there are fish and yabbies in the river, and cold beer and conversation at the Warrego Hotel. Fords Bridge consists of a crossroads, three houses and a pub, and when I arrive there's not a soul to be seen. Perhaps they're off along the Warrego, which lies brown and sluggish just outside of town. I venture into the hotel, a squat mud-brick building sheltering under a corrugated-iron roof. Out here, beyond Bourke, the pubs lie low to the earth, cowering from the sun, leaving high verandas and double-storeyed pretensions to the more clement climates of eastern towns. Inside, Andrea, the publican, is reading a book behind the bar. She seems mildly surprised to have a customer. I order a soft drink and she tells me Fords Bridge is experiencing a population explosion, growing from three to five.

'A friend of mine came out and there's a madwoman moved in down the road.'

'Who are the other two?' I ask.

'There's a retired shearer and another bloke', she says. She informs me that Fords Bridge has a lot going for it, including an 18-hole golf course: 'But only if you go round twice'.

I thank her for my drink and head back to the Bourke–Wanaaring road. It's dead straight, wide and well graded. Only once do I hit a boggy patch of red mud left by the rain, and even that's drying fast. Otherwise the road runs true under the wheels: driving is easy and there's plenty of opportunity to take in the immensity of the red land. This is grazing country, but I see more emus than I do sheep or cattle, and more goats than all the rest put together. Lizards scuttle across the dirt and a wedge-tailed eagle soars supreme on the morning thermals, its huge wings flashing yellow. The land is at once barren and brimming with life, and I stop to breathe it in.

To me it is magnificent; to Henry Lawson it was a living hell. In the summer of 1892 he tramped almost 200 kilometres through drought-ravaged land from Bourke to Hungerford and then back again. Looking out across the naked expanse, with the horizon melting into mirage, it's difficult to imagine, even in this good year of rain and relative plenty, how anyone could contemplate such a feat. Bush-walkers may clamber in their thousands across the Blue Mountains, pour over Kosciuszko and the Victorian Alps, and crawl all over the Tasmanian wilderness, but surely no-one would be mad enough to walk from Bourke to Hungerford. We love the outback, but we love it on four wheels with a boot full of provisions. Lawson recalled the trek in his poem 'The Paroo':

With blighted eyes and blistered feet,
With stomachs out of order,
Half mad with flies and dust and heat
We'd crossed the Queensland Border.
I longed to hear a stream go by
And see the circles quiver;
I longed to lie down and die
That night on Paroo River.

But when Lawson and his mate eventually made it to the Paroo, they found the river empty:

No place to camp—no spot of damp—
No moisture to be seen there;
If e'er there was it left no sign
That it had ever been there.
But ere the morn, with heart and soul
We'd cause to thank the Giver—
We found a muddy waterhole
Some ten miles down the river.

Lawson only survived the journey thanks to the generosity of a squatter, that and a few intermittent bores that dotted his route. Fourteen years before his trek, a bore sunk on a wing and a prayer at Kallara station near Bourke had changed the pattern of European settlement. The bore chanced upon a vast reservoir of underground water, the first artificial tapping of the Great Artesian Basin, that profound and mysterious pool of dark water that spreads out for mile after mile under Australia's northern and central deserts. To this day the underground water sustains people out here; when the rivers run dry and the rainwater tanks hold nothing but dust, the bores are the water of last resort. But the subterranean source is little understood, and no-one knows for certain how long it takes to renew itself. Some of its water is said to have lain deep in the foundations of the continent for more than a million years.

As the morning wears on and the exhilaration of the outback fades, I find new ways to entertain myself. Every 5 kilometres or so, a fence intersects the dead-straight road, leaving a narrow gap for traffic to pass over a cattle grid. The grids are raised above the level of the road. As I approach, I carefully line up the four-wheel drive with the gap and accelerate hard, ensuring I hit the grid at maximum speed, all the better to enjoy the momentary thrill of becoming airborne before the car slams back into the dirt. A wonderful thing, the rental car. Not that this is a sensible place to be playing silly buggers—have an accident out here and it could be hours before someone happens along.

After driving for three hours I reach Wanaaring, population 136; I've passed one vehicle during my three-hour journey. The first stop is the pub, the Outback Inn, another squat job, hunkered down low to soak up the coolness of the earth. There are no rooms available; they've been block-booked for a team of Telstra linesmen. But I'm not here for accommodation or beer; I'm here for the annual meeting of the Paroo River Association. The owner of the pub, Shiree, a dark-haired woman with a ready smile and a raucous voice, tells me the meeting is across the road in the local hall. The hall has no official name; it's just 'the Wanaaring hall'. No need for a name when it's the only one for a couple of hundred kilometres.

Soon the locals start arriving in four-wheel drives and farm trucks, locals being defined as anyone living within a radius of 150 kilometres. It's not a bad turnout, about twenty. The old bushman Banjo Burns and his ten-gallon hat have come up from Goorimpa, and Leon Zanker, his best blue shirt contrasting vividly with a sun-roasted face, has travelled more than 100 kilometres from Laurelvale. George Hackney, a recent arrival to the district, is in from the property he manages for the Sporting Shooters Association out on Wilcannia Road, and association president Rob Bartlett and his wife, Eveline, are there from Toonborough, loaded down with agenda items and reports. And it's a proper meeting, all right: last year's minutes are read, apologies received, correspondence recounted, and the finances laid bare. There are suggestions for improving the measurement of river height, ideas for raising funds through bumper stickers, and indignation at the curse of yabbie poachers. But these are side issues, being nothing compared to irrigation. For up in Queensland, on a property called Springvale, close to Eulo township, an upstart grazier named Jake Berghofer has started growing irrigated hay. For the flood-plain graziers gathered in the Wanaaring hall, this is anathema, a threat to their livelihoods and the river they hold dear. They know what's happened on the Culgoa and the other inside rivers; they know the Queensland Government is encouraging irrigation on the Warrego upstream from Fords Bridge; they know that water taken in Queensland means less for them. It's an internecine struggle: Berghofer is a fellow grazier, known to many of those gathered in the hall. He's referred to simply as 'Jake'. It's acknowledged that his plantings are small, but time and again the phrase 'the thin edge of the wedge' is deployed—what Jake has begun, he may expand; where Jake has ventured, others may follow. The meeting discusses legal strategies, political strategies and grass-roots strategies.

The Wanaaring graziers aren't without allies. Environmental scientists at the University of New South Wales are on side, concerned that the Paroo's southern lakes and their bounteous bird life may be deprived of life-giving floods. So, too, is the Environmental Defender's Office. Two young women lawyers have travelled from Sydney, their car now marooned outside the pub with a flat battery. One gets to her feet, outlining a strategy to mount a case against Jake Berghofer in Queensland's Land and Environment Court. Her inner-city inflections sound foreign inside the outback hall; the audience moves uncomfortably in its seats. These are outback graziers, not natural allies of greenies or lawyers or city dwellers. But the young lawyer persists, and soon they're listening to her words and not how she says them. She makes a convincing argument for taking court action, but doubts linger about the costs and possible liability. This is the outback: resolving disputes through the courts remains a novel concept.

By five o'clock the meeting is over and everyone adjourns to the pub. The beer flows and the real discussions start. Above the bar, a black umbrella hangs upside down from the roof. It's an unlikely sight, umbrellas being a rarity in these parts. Patrons are invited to chance their arm and flick their change inside, a task that becomes more difficult as the evening progresses. All proceeds go to the Flying Doctor Service. Out here, where every community and most properties have an airstrip, the Flying Doctor is spoken of with a fondness that borders upon reverence. Every pub has some device to extract donations: up at Eulo, patrons can ring a huge cow bell; down at Tilpa on the Darling, $2 will buy you the right to scribble your name on the corrugated iron walls like the thousands of others that have gone before you. A little further along from the unlikely umbrella hangs a dusty black negligee, its nether regions bulging with coins, also for the Flying Doctor. 'Don't ask', says Shiree. Her husband, Moc, a skinny bloke with a smile so big it has a hard time fitting on his face, has put in an appearance to help with the crowd. Their livewire son is there as well, back from the coast. The three of them keep up a constant flow of banter, chiacking their friends and customers. I get talking with a big, soft-eyed bloke called Andrew, his voice imbued with love and pride as he recounts the birth a few days earlier of his child in Brisbane. With his face scrubbed clean, a mop of black hair and a gentle manner, I take him for a schoolteacher or something similar. I'm wrong. He's a professional kangaroo shooter. Most nights Andrew's out spotlighting, killing the animals with single head shots before skinning them and packing the carcasses into a freezer to be sent off to Adelaide. He invites me out shooting the next night, as if for a round of golf, but I make my excuses that I need to be far away.

The voices around the bar are growing louder as night falls and the beer starts doing its work. Shot glasses of rum start to cross the bar. In the corner, Banjo Burns, ten-gallon hat tilted back on his head, is telling a yarn about a horse with a stockwhip stuck up its arse, much to the delight of those listening. Behind me, a less gifted raconteur is repeating a threadbare joke disparaging the numeracy of Aboriginals. Around the room, small groups are passionately discussing everything from cattle prices to weather to local deaths. Then Shiree reappears, silencing everyone with her booming voice; she's cooked up a feast on the barbecue out the back, and everyone's invited, on the house. The crowd protests, insisting on paying, but Shiree is having none of it. 'You have us for dinner, so why can't we have you lot?' It's local steak, thick and juicy, served with coleslaw and tinned beetroot. I sit out under the stars, balancing a plate on my knees and talking to one of the Telstra linesmen who has emerged from his room at the prospect of a free feed.

Returning to the bar, a few people have drifted away but a hard core has settled in and the alcohol is loosening tongues. It seems Banjo has been done by fishing inspectors for having a net too many in the river. 'I was just getting a feed, for Pete's sake', he says, repeating the time-honoured defence.

Leon is indignant on his mate's behalf. 'You've got these bloody gangs coming up from Mildura, ripping off truckloads of yabbies, and these bloody inspectors are targeting the blokes who can help catch 'em. Well stuff 'em, they're not coming back on my place.'

'You can't stop 'em', says Rob.

'Well I don't have to help 'em.'

The conversation, inevitably, has returned to the river. Paroo yabbies are a delicacy, fetching $25 a kilo or more down in Mildura, or so it's said. It strikes me as a somewhat unlikely destination for such exotic contraband, but what would I know. The group around the bar discusses the merits of whether or not to let fishermen onto their land and under what conditions. Some close off access altogether; most are happy to have friends and friends of friends come on. One bloke is more welcoming. 'I want people to come on, you know I do. They're all welcome. But I won't have the river raped. I won't have it.'

And from yabbie pirates and incompetent fishing inspectors, it's just a small move back to Jake Berghofer and irrigation, to lawyers, scientists and politicians. 'Dynamite was a lot easier', says Rob, and heads nod in agreement, some smiling, some not. Later, I ask Leon about the dynamite. He reckons that it's not the government that has kept the river flowing all these years.

'There's heaps of old banks put through this country', Leon explains. 'Banks to divert the floods. There's banks on Rob's place, Toonborough; there's banks on Goorimpa. There were banks on Nocoleche, which is a national park now, including a bank that went across the main channel. And that was holding back shitloads of water from coming down. So blokes around here found out about it and said that's not on. So they went up there one night with shovels and gelignite and blew the shit out of it. Blew it up. They took the law into their own hands. "Bugger you", they said. "You're not stopping us getting our water. It's no more your water than it is mine."' In my head I do the calculations, visualising the map I've been studying for days. I know where Nocoleche is, and roughly where the dozen or so downriver properties lie. I look around the bar with new respect. It's the last wild river in the Murray-Darling all right.

But it's getting late, closing in on midnight, and it's time to call it a night. Leon and Banjo stock up with two or three stubbies each for the long drive home, over 100 kilometres along deserted roads. But times are changing in the outback: Leon's wife is the designated driver. As the pub empties, Shiree takes pity on me. I'd been intending to pitch my tent by the banks of the magical Paroo, less than 100 metres away across the road, but somehow I haven't been able to get away from the pub during daylight hours. Now, the prospect of setting camp in the middle of the night, on unfamiliar ground, with a belly full of beer, doesn't look so promising. Instead, Shiree gives me the keys to the Country Women's Association house across the road. 'Twenty-five bucks. All proceeds to the CWA', she says. I nod gratefully and stagger off to get my gear from the car.

⌒

I wake at dawn with a dry mouth and a belligerent head. I totter down the twilight corridor of the CWA house and brew a strong coffee in its spartan kitchen. Judging by the empty bourbon bottles in the bin, I'm not the first traveller to have found refuge in the old house. Either that or the CWA is not the organisation it once was. Nursing my mug of cafe-au-long-life-milk and a two-day-old pastry from Bourke, I go out the back door and walk barefoot across the dewy ground to the river, down where I'd intended pitching my tent. The morning is still and cool, with the sunlight glancing low through the trees and across the water. I sit by the semi-exposed roots of a river red gum and behold the tableau. The surface shimmers ever so slightly with the first hesitant zephyrs of a new day. The river isn't full within its shallow banks and it's barely flowing, yet it's a good 10 metres wide and winds its way uninterrupted in both directions. There's no sign of the red duckweed that clogs the weir pools of the Darling. The river red gums haphazardly line the shallow banks, their branches strong and their foliage full. A cockatoo, aglow in the early rays, alights on a treetop, white against green, but refrains from its signature screech, respectful of the stillness. As if on cue a fish leaps and across the surface circles quiver. If only Henry Lawson could see the Paroo as I see it this morning: near enough full, with life abounding.

I contemplate the river's journey. There's something satisfying in the knowledge that it's arrived here of its own accord: not depleted like the Culgoa; not constrained like the Balonne; not pumped and primed and piped and regulated and rationed and bought and sold and bartered like all those other rivers of the Murray-Darling. This water carries no uncertain provenance: it is here because some months ago rain chanced to fall in Queensland and the water made its way unimpeded to this spot, assisted by nothing more than gravity. A duck wheels in and lands on the water before paddling self-importantly off towards the far bank, fossicking for food. I sit breathing in the scene well after I've finished my coffee. Eventually, I return to the house to brew another, but the signs are good: the hangover is in retreat.

⌒

I drive south, roughly parallel to the Paroo along the western edge of the flood plain. No clouds again today, although here and there the evidence of the previous week's rainfall persists in puddles beside the road. The earth is bright red, dust and rock, with a low covering of mulga scrub. I reach Nocoleche Nature Reserve, where two pink-and-white cockatoos perch in the branches of a dead gum to welcome me. The inevitable goats scamper across the road, and a wedge-tailed eagle floats just metres above the road, practising on the day's nascent thermals. A family of western red kangaroos, safe within the reserve from soft-eyed marksmen, lies in the dubious shade of a mulga tree, a joey peering curiously from its mother's pouch. A fat lizard scurries away before I can train my camera on it.

Fifty kilometres south of Wanaaring I reach Rob Bartlett's property, Toonborough. The sign at the turnoff is a big old circular-saw blade painted white, with the name in black. Once onto the property, I find that the track zigzags through the black-soil scrub. A flash of light catches my eye: sunlight on water, a lot of water, glimpsed through the trees. Rob meets me at the house and we pile into his truck. A canny man in his late fifties, blue eyes alert behind his wire-rim glasses, Rob is looking none the worse for an evening at the pub. In a moment we're back at the water, a glistening natural lake, hundreds of metres across. The ghost of Henry Lawson groans at its bounty. We climb out to appreciate the spectacle. The lake, like the river, has receded from its peak, and 10 or 20 metres of lakebed are exposed around its edge. Sheep are chewing at the green shoots sprouting from the dark soil.

'This in Mungundi Lake', Rob says. 'It fills from the Paroo. We've been here forty years and it's only gone dry three times, once for six months. Other times it rises and falls depending on what comes down the river. There are not many birds at the moment, but we've had four or five hundred ducks on it at a time, whistler ducks and others. Plus pelicans and shags.' If Rob is apologising for an absence of bird life, he needn't bother. There are ducks on the lake, parrots in the trees, and an ibis is wading along the shoreline. Bird calls fill the otherwise silent morning.

I ask Rob if he's ever been tempted to irrigate. 'No, not really', he replies. 'The thought went through my mind probably thirty or forty years ago. But it's not sustainable on the Paroo. It's not a regular supply of water. If we run out of water here there's no water for stock and the gardens and the house. The house is supplied from this lake. And you have to live in the area as well. Our attitude to water is that everyone should have their share, including nature. It should flow through, it's yours when it's on your country and while it's flowing through, but once it's gone, it's down onto another place doing a bit of good down there.'

Looking out across the expanse of water, I say it's hard to believe this is the worst drought in a century. Rob shrugs and tells me that out here it's not. He reckons the 1940s drought, when the Darling stopped five years out of six at his parents' property near Tilpa, and the one in the 1960s, including here on the Paroo, were far worse. 'The Paroo in the past seven years, when it's supposed to be the worst drought in a hundred years, we've had three or four major floods. It's been the only river in New South Wales that has reached its peaks.'

Rob has a theory about the current drought, about how it affects the river system. 'The Paroo is pretty much as it always was: no extractions, not much water taken out of the river. If you look at the inside rivers, every little place has got its own dam or two dams or three dams. I've seen 25-acre blocks around Dubbo that have four or five dams on them. They're all contour furrowing, so they're getting better at holding every bit of moisture on their land. Which is good farming practice, but that water should have been running off. It isn't getting into the little creek, so the little creek isn't getting into the river. It's just not happening. We're not ever going to see the rivers run like they used to because we're getting too good at holding the water inside farming country.'

Back in the truck, Rob tells me he bought Toonborough back in 1969. He's not sure where the property got its name. Like sailors with boats, graziers don't like changing the names of the properties they buy. Rob is a third-generation grazier, lured to the Paroo from the Darling by the promise of flooded country. Toonborough covers 28 000 hectares. Rob's son Ben has the place next door, another 41 000 hectares. Add some land leased across the river and the two men are grazing around 80 000 hectares—80 000 hectares to carry about 600 breeding cows and 5000 or so sheep, with only contract shearers and crutchers to help. River or no river, it's not high-intensity agriculture. The truck accelerates across a claypan, then slows to push through lignum bushes to reach the river proper. Another revelation. It's almost another lake, wide and plentiful, if again below its peak. Our arrival startles a squadron of pelicans that has congregated mid-stream. They use the breadth of the river to climb ponderously into the air. I wonder what Henry Lawson would think of pelicans. 'This is the main channel of the Paroo', says Rob. 'It's not a permanent waterhole, but it's a good waterhole.'

Down along the bank, in the shade of a river red gum, lies a grave, adorned on either side by two unopened cans of Victoria Bitter, their metallic green turned a powdery blue by the passing of time. 'John

"Macca" MacDonald 24.10.2006' reads the inscription. Rob says it's not actually a grave, rather a memorial to the Mildura fisherman who died of a heart attack at this spot one spring evening. The next year his mates came back for a wake, erected the memorial and left the beers behind. 'He apparently liked a drink', observes Rob. 'He's been christened a few times by the dogs. No disrespect to him, of course. The tree he's under is at least 100 years old, probably a lot more. Not a bad spot, when you think about it.'

Rob reckons the long periods of separation from the Darling have encouraged unique species of fish and yabbies to prosper despite the arrival of European carp. The hated carp arrived with the massive floods of the mid-1970s. The Paroo washed through its lower lakes and its previously untainted floodwaters intermingled with those of the Darling, allowing the carp to swim up against the flow and infect the Paroo like some piscine venereal disease. Like everywhere else, the arrival of the vermin was greeted with despair, but Rob says that in recent years the carp have been in decline. 'That lake at the house went dry in 1980. There were thousands of dead carp. It stank for the whole six months it was empty ... It went dry again in 2003, and there were only about three or four carp out there. We don't know what is happening, but for whatever reasons, the native fish are making a comeback.'

We return to the house for a revitalising cup of tea. Rob's wife, Eveline—'just call me Ev'—recounts how the advent of electricity here in 1990, and more recently the internet, have made life easier. She also recounts floods with affection. The people here recall the dates of flood years like they do the birth years of their children, recording river heights with lines on gates and outbuildings as they do the heights of growing offspring on kitchen door jambs. Ev recalls the time when two wool classers were marooned on the property by the rising waters.

'One bloke couldn't put up with it; he was going stir-crazy. So he set off after dinner one night to walk to Wanaaring. Fifty kilometres. The water's over the road and it's pitch black. All night and all day he walked. Wading through the water, full of mosquitoes and who knows what. The other bloke said, "I'm not going to do that". So he waited. And sure enough, a helicopter came by and picked him up. And just as they were landing in Wanaaring, who should emerge from the bush but the first bloke, after having walked for a day and a half.'

Another 60 kilometres down the road and Leon Zanker is waiting for me, working in his machinery shed as I pull into Laurelvale. He welcomes me with a bone-crusher handshake. His face is open and hearty, a wide smile revealing uneven teeth, and grey-tinged sideburns crawling down beside sunburnt ears. He's held off eating lunch until I arrive. We go inside the house and he extracts the makings from the fridge: a cooked chicken, iceberg lettuce, sliced cheddar, canned beetroot and wholegrain bread, to be washed down with bright-green cordial. We talk as we construct sandwiches and Leon describes just how sensitive the river is this far south to any human interventions. He explains that Laurelvale lies on a flood plain with no real river; the land is so flat that the main channel has fragmented.

'From Toonborough down it stops being a river; it breaks out and really trickles down. We had a little trickle in 2004, which was good at Rob's but not here. It's frustrating to see Banjo at Goorimpa get a lovely little flood and then it stops at my boundary. He's got fat cattle and mine are being trucked out. Next year it comes into my place for miles and I get a big flood and I'm laughing … the really good flood country starts from here down. When you get down the bottom, it's black, spewy, flood plain-type of country. Very friable, very rich. I've never seen fatter cattle. There's maybe ten places under here that get good flooding. You're talking a million, a million and a half acres. It's such a shallow fall, a couple more inches can mean a huge spread. It's like a fan. That's why we're so anti anyone touching anything up the top. If you take the guts out of that flood surge, that equates to thousands and thousands of acres down here that won't get covered.'

While demolishing his sandwich, Leon explains that simply by building a bank no higher than the table we're seated at, he could cut off one of his neighbours. But the advent of the internet and satellite imagery has changed the ground rules. Late into the night, graziers are on Google Earth, searching for the straight lines that betray channels and banks. And so are the scientists and environmentalists in Sydney. And all of them are watching what Jake Berghofer is doing in Queensland.

'The capacity that Jake takes is insignificant', says Leon. 'But that's not the point. Because if you let one operator do it, then if he can do it, why not I. And Jake argues "I'm not using hardly any water". But that's what they said on all those other rivers: the Warrego, the Culgoa, the Gwydir, all the rivers that come down into the Murray-Darling Basin. They all started with one irrigation development too. And what are they now? Totally overallocated and completely stuffed. And now there's no water getting down as far as Lake Alexandrina and Lake Albert in South Australia and it's all back on the river systems. Every river system is fucked. Now we've got a government that is saying I'm going to use your taxpayers' money and I'm going to buy back the water. Not thousands of dollars. Not millions. Billions. Billions. And they've only just started. I guarantee that when you add it all up, add up all the income made out of irrigation, and then you add up all the cost and repairing all the damage done, in five or ten years time you tell me which is in front?'

From Laurelvale I continue east, back towards the Darling. I cross the Paroo flood plain, but the main channel, so clear and obvious just 60 kilometres north at Toonborough, has devolved into scattered channels. In the late afternoon, after traversing some low-lying ridges, I arrive at Tilpa on the banks of the Darling. I call in at the pub, handing over my $2 for the Flying Doctor and writing my name on the wall. I take my drink out the back through the beer garden to the banks of the Darling. The river channel is so much bigger, so much deeper and wider than the Paroo, carved deep into the plain instead of running over the top of it. But it's a channel that is all but empty, the water not moving. On top of the banks, the river red gums are alive, but many carry dead branches. Tilpa, population six, had better be careful: Fords Bridge may soon overtake it. I leave the pub, driving north up the western side of the river towards Louth and Bourke. The flood plain is grey: grey dust, grey trees, grey grass. The Darling at sunset is a very different river from the Paroo at dawn.

Cunnamulla, they say, is a Cunnamulla of a place. The town lies on the Warrego River 250 kilometres north of Bourke, across the Queensland border. It's immortalised in the Slim Dusty song 'Cunnamulla Fella' and in the Dennis O'Rourke documentary *Cunnamulla*. Slim Dusty sang:

Now I've done a little fightin' in the western bars,
Done a little lovin' 'neath the moon and stars
I wear bright clothes and shirts full of colour
And the girls know me as that certain fella
I'm the fella from Cunnamulla
I'm the Cunnamulla fella …

O'Rourke, on the other hand, outraged the local citizenry by focusing his camera on a select collection of misfits, depicting the town as insular, dysfunctional and racist. In honour of Slim and the song, a twice life-sized bronze statue of a square-jawed, clear-eyed bushman, the Cunnamulla Fella, manufactured by a sculptor of note in Grapevine, Texas, graces the town park outside the council chambers. As far as I can ascertain, no monument, lasting or otherwise, has been erected in honour of O'Rourke.

I find Jake Berghofer at the town weir, waterskiing with friends. 'See. Put in a proper weir and everyone wins', he says above the roar of an outboard engine, the hint of a smile on his lips. He's dealt with journalists before and doesn't mind baiting them. Down on the water, four beefy blokes with their shirts off are manoeuvring the powerboat to shore past the skeleton of a tree drowned when the weir was erected in 1989. Jake fishes an ice-cold can from a nearby Esky and hands it to me. We sit in the shade of a purpose-built platform beside the river. We're on a block of land owned by one of Jake's mates—a mate who owns an earth-moving business and is growing prosperous building farm dams.

Jake's a young bloke, just thirty-one and already starting to develop a middle-age spread. He's friendly and earnest, with a larrikin vein just below the surface. He's wearing board shorts, a yellow 'cat racing' polo shirt, sunglasses and a green cap embroidered with the name of his property, Springvale. Jake has come into Cunnamulla for the weekend. He lives out near Eulo, 70 kilometres to the west, a stone's throw in the outback. The tale he tells reveals an ambitious young man. When his family moved from Toowoomba and bought the Eulo store, Jake was just seventeen. His family hunted around for a property nearby to set him up and settled on a small place, just 16 000 hectares, called Springvale. The way Jake tells it, he didn't have a clue about grazing. 'No. I'd never even ridden a motorbike or a horse or nothing', he laughs. 'We employed an Aboriginal guy who had worked on the property for a few years. He was the first one to teach me the ropes. Then I'd work for neighbours for free. Instead of being paid, I'd help them out. Then they'd come and help me out when I needed it. It's a great community.' While we talk, one of those neighbours joins us, Randall Newsham. Randall looks about sixty, but clearly likes and respects Jake. Jake may be viewed as the devil incarnate south of the border, but he has plenty of support in his own community.

The way Jake tells it, the Berghofers were attracted to Springvale for two reasons: it was close to Eulo and it had irrigation permits, not to pump water from the Paroo but to collect and store water that ran across the property. 'It was part of the advertising of the property that it had an irrigation licence. It was all licensed back in the sixties. And that is where a lot of the misrepresentation from New South Wales has come from. They say I've done all this in the past few years. Well that's just not true. I've got all the original licences that aren't even in my name. It was done in the 1960s … and there was irrigation on the property back then. All the hay-making gear and an old header.'

Storing water that flows across your property is known as flood plain harvesting and is anathema to the likes of Rob Bartlett and Leon Zanker. But Jake reckons his situation is unique. He maintains he isn't harvesting floodwaters, nor is he preventing rainwater from reaching the Paroo. 'I've got a large 1000-acre claypan, which is solid clay. It doesn't grow anything. Even when it has water in it, you can drive a truck over it, it is that hard … So if you get an inch of rain, you get an inch of water over 1000 acres. I've got it designed with little channels in it that takes it into 10 acres. So that 1000 acres would evaporate after about a week, so I put it into 10 acres, which makes it last a month. So if I get 4 inches of rain, I get 200 or 300 megalitres, which is enough for me to grow 100 acres of hay.'

Jake says his farm practice is environmentally sound because he sells all his hay locally. Any hay his neighbours truck in comes hundreds of kilometres and leaves a significant carbon footprint. But he doesn't pretend it's not a nice little earner. 'I can make more off that 100 acres than the rest of the 40 000. I could shut down the rest of the property and just live off that and still be fairly viable.' Jake invites me out to his place to have a look around for myself. I accept. He's staying in Cunnamulla for the night but I decide to keep going, to head out to Eulo and see what the Paroo looks like in Queensland.

In New South Wales, Bourke is the high water mark of the push to develop the Western Division. It's where the bitumen ends and the railway line finished. Cunnamulla, its northern equivalent, still has rail freight, and the road into the outback is sealed for 300 kilometres further west. If Sydney has been retreating from the bush, then Brisbane has continued to push westward. The weir at Cunnamulla was built fifteen years after St George's was completed.

Perhaps it's the sealed road that makes the trip out to Eulo lack the drama of the trip between Bourke and Wanaaring. Perhaps it's because the country is in slightly better condition, with higher rainfall. Perhaps I'm just becoming inured to the flat plain. But the thin strip of asphalt can't disguise Eulo's identity as an outback town. 'Welcome to Eulo. Lizard Country' reads the sign at the edge of town. 'Population ... 50 people and 1500 lizards'. A 3-metre-long frill-necked lizard, a papier-mâché-like construction of wire and concrete, looks out across the town from its perch atop a derelict windmill tower. It's there to guard the lizard-racing track, a nondescript patch of red earth in a vacant lot. For three days in September the town holds lizard races, an excuse for people to come in from far and wide, drink a lot of beer, and pack in enough socialising to see them through to Christmas. Most outback towns host similar events for similar reasons, from the celebrated picnic races at Birdsville to the annual goat muster at Wanaaring: a chance for locals to catch up and rope in a few elusive tourist dollars at the same time. I'm told that a few years ago the lizard races were thrown into turmoil when the reigning champion, Woodhead, was beaten by a 'cockroach' from New South Wales called Destructo. Destructo was unable to defend his title; he was 'accidentally' trodden on shortly after his victory. The rivalry between Queensland and New South Wales runs deeper than rugby league and disputes over water.

At the Eulo Queen Hotel, the publican, Bill Prentice, is worried about water. He has two pot glasses of it sitting on the bar when I enter. It's murky brown, with a half-inch of sediment lurking in the bottom. 'Here, smell this', he says, not bothering with introductions, and hands me one of the glasses. The water looks like sewage but reeks of chlorine. 'Would you drink that?'

'No thanks', I reply. 'What is it?'

'That, my friend, is the Eulo water supply.'

Eulo is far enough north to tap into the Great Artesian Basin. A bore was sunk in 1966 and has been supplying the town with water ever since. But now, Bill tells me, some interfering health bureaucrats have decided to chlorinate the water. He indicates the glasses on the bar: exhibit A, more evidence testifying to capital city incompetence. I offer no arguments and settle in for a beer. The Eulo Queen, named after the sly-grogger, publican and brothel madam Isobel Gray, who ran Eulo during the opal rush of the late nineteenth century, conforms to the low-slung, ground-hugging outback pub style. Behind the bar, with its corrugated-iron facade, the woodwork has been scorched with the cattle brands of local grazing properties. I rent a cabin out the back and order the pub's signature dish for dinner: a plate of Cunnamulla sausages.

The next day, while I wait for Jake to show up, I take a stroll around Eulo. It doesn't take long. First stop is the lizard lounge, a picnic area across the main road from the pub and racetrack. It's constructed from corrugated iron, with steel seating and shade cloth. There's even a smattering of green grass on the red earth. There's no sign of the lounge ever having been used, and it most probably isn't, except during the three-day spring festival. However, a sign sets out a map of the Eulo Heritage Trail, listing the town's six highlights: 1) the bore, 2) the pub, 3) the shop, 4) the air-raid shelter, 5) the post office, and 6) the lizard racetrack. I hope Jake shows up soon. I head off to find the air-raid shelter and locate it in the backyard of the shop: a 6-metre-long trench covered by curved sheets of corrugated iron. A heritage sign informs me it was built as a bomb shelter during World War II and has 'stood the test of time'. At a pinch, it would still hold half the town's inhabitants, minus the lizards. Inside, a couple of empty bottles and a pair of discarded boots lie on the dirt floor. I admire the rudimentary architecture of the shelter, and the logic behind it even more: first Singapore, then Darwin and finally Eulo: the Japanese invasion plan revealed.

Inside the shop, there is no-one behind the counter. It takes a few yells before a portly chap in his fifties shuffles out from a back room, his long white goatee contrasting with his still-dark hair. A pair of reading glasses borrowed from the stock perches forgotten on his wide head, their magnification identified as 2.50 by a small white sticker on one of the lenses. We get talking, polite small talk, until he learns I'm writing a book about the river.

'Water, hey?' he says, suddenly engaged. 'You don't want to know the real story though, do you?'

'Of course.'

'I doubt it. They wouldn't let you write it anyway.'

'Who wouldn't?'

'You'd better come out here.'

He turns and walks through a doorway to the back of the shop. I skirt the counter and follow him into the office. It's a shambles. Cartons of stock fill one corner, opals and polishing equipment clutter a second, while in a third, a computer protrudes above a desk awash with papers. The storekeeper tells me his name is Garry, that he'd like to sell the store, that he wants to concentrate on buying and selling opals. We start talking opals and, as if attracted by the mere word, an old codger emerges from somewhere deeper in the building. He's as skinny as a rake, sultana-shrivelled by the outback sun, and wearing clothes that have the appearance of being freshly washed and dusty at the same time. He introduces himself as Don Murphy, aged seventy-nine, an opal fossicker by profession and choice. 'I retired back in 1997, gave the property to my sons. Thought it would be good. But all I did was sit around on my bum in Charleville eating ice-cream. I had a heart attack. "Bugger that," I said, "I'm not waiting round here to die". So I came fossicking and been here ever since.' Don works with two others, working a series of 2-metre-deep holes out in the desert. He won't say exactly where. He's coming to the end of the season; it's too hot to work through summer. 'There's money in opals', he confides in a low voice, and his eyes gleam.

Once Don heads off, Garry returns to water.

'Do you know what a contrail is?'

'You mean from aeroplanes?'

'Yep. We call 'em vapour trails but the yanks call 'em contrails.'

There's an awkward moment. Decades of journalism warn me where this is going, while Garry appears to be pondering whether or not I can be trusted. After all, I am from Canberra. But the truth must out.

'Ever heard of Monsanto?' he asks.

'Big agrochemical company. Makes genetically modified crops.'

'Smart chap. Well, Monsanto also controls the contrails. You know why?'

'Don't tell me. World domination?'

But the sarcasm is too little, too late; Garry needs to tell me what he knows. 'Water. They're using the contrails to dry out the planet, to change the climate. It's nothing to do with burning coal or driving cars. They're using a science called "chemics" to dry everything out. They want to control water, so it will be worth more. The internet is full of this stuff. You need to check it out. I spend my nights back here, polishing opals and surfing the Net. Just google "contrails", you'll see what I mean. Thirty years ago, there was plenty of water in the rivers, before the contrails. Now, if you hear a plane, you know there won't be rain.'

The conspiracy widens, then deepens, then gushes. It's joined by another tributary. Garry has uncovered the cause of cancer: acid water.

A pH below 4.5 and cancer is inevitable. 'At the moment, one out of sixteen people dies of cancer; they want to get it down to one in two.' He says the answer is alkaline water. Natural rainwater has a pH of 6.5, no lower than 5. Bore water is better, very alkaline, up to a pH of 9. Drink alkaline water and you'll be okay. I ask if that's why he lives out west. 'One reason, but have you seen what they've done to our bore? Worked perfectly well for forty years, now they're trying to stop us drinking it. They don't like people living out here. We're too independent; we can get our own food and water ... We're at war here. A handful of people have control over the world. The government doesn't have much choice; Nestlé is eight times the size of Australia.'

And on it goes. Eventually I extract myself from the confused room and return to the harsh sunlight of reality. Somehow the 65-year-old air-raid shelter no longer seems so out of place. Round on the shop's front veranda, a pallet of blue cardboard casks of alkaline water are on sale. They look like they've been there for a while, but perhaps the sorry state of the bore will rectify that. I'm about to wander off when the penny drops. I walk back into the shop and through to the office.

'Hey Garry. How long have you been here?'

'Too long. About fifteen years. Why?'

'Are you related to Jake Berghofer?'

'Sure. I'm his dad. Why?'

'Nothing. Just wondering. I'm catching up with him later on. He's going to show me his hayfield.'

'He's in Cunnamulla. Back tonight.'

⌒

Brian Leuthford is a beekeeper. There are three or four based in Eulo, but Brian is the only one who doesn't take his hives on the road over summer. Beekeepers, like roo shooters, are welcome on local properties. They'll let the graziers know if gates have been left open, fences are down, or stock are injured or bogged. And there's always a pot or two of honey to be had. Brian has lived in Eulo for sixty years. He's a gentle man with a gentle profession, still intrigued by the insects that provide him with a living. 'They're a lot smarter than they look. You never know everything about them. You think you know something about them, but if you knew what you don't know, then you would know a lot more than you do know.'

The beekeepers come for the yapunyah trees that grow on the flood plains of the outside rivers: the Warrego, the Paroo, the Wilson and the Bulloo. The yapunyah flowers throughout winter, from April to November, providing the bees with a constant diet of pollen and nectar. 'There were probably 40 000 beehives along the Paroo during the winters in the seventies, after the big floods. They came from as far off as Brisbane and Melbourne. Even so, you'd still see the nectar dripping from the trees and going onto the ground, there was so much. It was a good thing', says Brian. Nowadays, there are maybe 10 000 hives on the Paroo over winter. Depending on the dryness of the flood plain, Brian reckons his 400 hives can produce anything from nothing to tonnes of honey in a year.

'You'll see all sorts of things. You're in the bush, all alone. All the birds and animals. It used to be better before there were u.h.f. radios and mobile phones. You're out by yourself. Nobody can get hold of you. You might go out for two or three days. Just take a swag and camp out there. You see the night and the day, just lying there looking at the stars. It's a good thing.'

Brian's livelihood depends on the yapunyah trees, and they in turn rely on the health of the flood plain. Yet when I ask him if he is concerned about Jake Berghofer and his irrigation scheme, Brian's support for his neighbour is unequivocal. 'It's our water. It's fallen here, the rain. Why should we have to send it to New South Wales? They want to use it, so why can't we use a little bit on the way? It's not new, you know. I travel around these old properties; there are old mowers and chaff cutters and pumps out there in the scrub. And he doesn't pump out of the river, so it's not affecting the river at all. He's taking water out of a claypan. Otherwise it just sits there and evaporates. It's a good thing.'

⌒

Jake bowls into the pub in the early evening, full of apologies for being late. We get going while it's still light, Jake leading in his powerful four-wheel drive as we head across the weir and out of town, into the setting sun. But once we turn south onto dirt, he leaves me behind in a cloud of blood-red dust. I miss the turn-off to Springvale, and only the receding dust ball heading over a rise guides me as I retrace my way. At the house I climb into his truck and we bash across the hard red ground of his property. It's not exactly hilly, but the land has more rise and fall than any other property I've seen this far west. In no time we're over a low ridge and arrive at the claypan. It's big, about 400 hectares, roughly 500 soccer pitches. Jake has bulldozed a 2-metre-high earth levee around it, with a series of channels in the clay bed to run the water off towards a holding dam. On top of the levee sits an old truck, its cabin tilted forward, exposing the engine. The drive shaft has been connected to a pump, an ingenious piece of bush engineering. The pump takes the water out of the claypan and sends it through a canal to the field where Jakes grows his hay. 'You see, it's all self-contained. The water would never reach the river', says Jake proudly.

It's evident the work is recent, not more than a couple of years old, and this has me intrigued. That's because in 2002 the Queensland Government declared a statewide moratorium on irrigation. The next year, then Premier Peter Beattie and his New South Wales counterpart, Bob Carr, travelled to Hungerford, rowed out onto the river and announced the Paroo River Agreement, securing the river's future, or so it was thought. Back in the truck we drive cross-country towards his hayfield and I ask him how it was possible to build new works despite the moratorium and the Paroo River Agreement. Jake answers that his scheme is not new, not technically. There were works on Springvale before he bought it, and all he had to do was notify the Queensland authorities that he intended to reactivate the works. By doing so, he was allowed to restore and renovate those works.

'So tell me', I ask. 'Is this it, the claypan and the hayfield, or are there other pre-existing works you could revive?'

'I do have some storages that could take water out of creeks—not the Paroo. But at the moment, to date, the only water I use is about 5 kilometres from the river.'

'But you could build, or rather rebuild, other storages?'

'Yeah, but not so much to store it as to spread it.'

'What does that mean?'

'Well, if you have a creek running and they put a bank across the creek and the water comes out like this and floods all this country and then comes back into the creek, it makes all this country wet ... so it grows more feed. So it's not so much holding it back as spreading it out here.'

'So it's expanding the flood plain?'

'Yeah, basically, but ... the department still considers that as irrigation.'

And if Jake wanted to, he's entitled to increase his take of water manyfold. At present he takes several hundred megalitres a year from the claypan, but he's licensed to take thirty to fifty times as much from the flood plain, thanks to those pre-existing licences. Ironically, he's being held back by another moratorium, one on clearing land; his property is covered in scrub and he isn't permitted to chop down the trees to create fields. So instead, he's lobbying to sell his water rights to neighbours like Randall Newsham, who owns land that is already cleared and where the soil is better. It's not a pipedream either. The transfer of the licence would not break the moratorium established by the Queensland Government, which has happily given the green light to the sale and reactivation of sleeper licences over on the Warrego.

And Jake is not alone. I recount what the beekeeper, Brian Leuthford, told me about the old properties, with their derelict pumps and hay mowers. Is it possible that others will reactivate their rights and rebuild their storages?

'Well, off the record ... No, it doesn't have to be off the record. The department knows now. There are a few that haven't notified yet. They're not really irrigators, it's more just ponding, for pasture and that sort of thing ... It's not too late for them. They can still do it, but people don't know they exist because they haven't been through the process I have and made myself legal.'

We reach the hayfield, a large circular paddock, much smaller than the claypan that nourishes it. The hay has recently been harvested and in the fading light it isn't much to look at. Anywhere else it would be entirely unremarkable, but on the Paroo these 40 hectares excite passions on both sides of the border. It's almost dark by the time we get back to the house. Jake apologises again for getting back to Eulo so late in the day, and I thank him for the tour. But driving back through the scrub to Eulo I conclude that the graziers down at Wanaaring are right: Jake is the thin end, and it could be a bloody big wedge.

The pub that evening is unexpectedly busy. The three grazing families that stopped for lunch are still drinking on the veranda, the kids tucking into pies and fish and chips in front of a television inside as the adults grow increasingly mellow outside. Denny, the former owner of the Eulo Queen, is in fine form. He's swaying over the pool table, so pissed he's having difficulty focusing on the cue ball, let alone hitting it.

'He looks a bit worse for wear', I say to Bill as the publican pulls me a beer.

'More like falling down drunk', says Bill, shaking his head. Denny immediately vindicates Bill's analysis, landing face-first on the wooden boards with a soft splat, contrasted by a sharp report as his pool cue strikes the floor beside him.

'Oh shit', sighs Bill, rolling his eyes as he comes round the end of the bar to give me a hand getting Denny upright.

'D'you see that?' slurs Denny, stinking of beer, as we get him vertical. 'I fell down. Jus' like that. Plop.' And he imitates the motion with his hand. We help him over to the bar. 'I'd better have a rum', he says. Bill shakes his head to himself again, but pours the shot glass of Bundy just the same. A few hours later, as I head off to my cabin, Denny is still going strong, out on the veranda chatting with the graziers, appearing none the worse for wear.

Before I leave Eulo, I go out along the road to the weir to pay my respects to the Paroo one last time. The weir is barely worth the name, sitting no more than a metre high within the channel. But in these flatlands it's enough to back the water up 5 kilometres. And on the lee side of the weir the river is all but empty, just a couple of disconnected waterholes baking in the sun. The observation is meaningless: without the weir, the whole river would be a series of disconnected water-holes waiting for the next surge of water from the tin roof country to connect them. I like Jake. He's trying to make something of Springvale. But I think he's wrong. Not because his arguments are wrong, but simply because he is arguing them in the wrong place.

The Paroo is the last wild river, flourishing in defiance of drought, governments and contrails. It's worth preserving, not for the sake of the Wanaaring graziers, but for its own sake. And I think it will be. Sooner or later some smart political operative is going to make the same calculation that Peter Beattie and Bob Carr made: that moving to save the Paroo is a lot easier, a lot cheaper, and will put a lot less people offside than addressing some of the intractable problems elsewhere in the system. They'll save the Paroo all right, a fig leaf to cover whatever they won't or can't save elsewhere. But who cares what motivates them. None of the other rivers are ever again going to approximate their natural state. But out on the western plain, preserved until now by its remoteness, intermittent flows and a few well-placed sticks of industrial explosives, flows a real river; shallow and ill-defined to be sure, but a river as nature shaped it. To borrow a phrase from an old beekeeper, it's a good thing.

4

AN ANCIENT LAND

Lake Menindee: seven years empty

It's been raining again, and the lawn has obstinately refused to die despite the advent of summer proper. The long absent call of the motor mower has returned to punctuate the calm of Canberra's suburban evenings. Like dormant seeds or migratory birds, the machines have never truly vanished; they've simply been waiting for the rains to return. Perhaps the drought is breaking. The local authorities certainly seem more confident of this, or maybe they've just become more practised at crisis management. In their magnanimity, between mid-December and the end of January they're permitting the use of sprinklers one day a week—between seven and ten o'clock on Saturday night for those of us with even street numbers, and between the same hours on Sunday evening for those with odd numbers. It hardly seems worth it if, come the heat of February, the lawn is simply going to wither away anyway. Still, I might get that old sprinkler out from under the house and set it going one balmy evening so the kids can run under it, just so they know how it feels. If the lawn gets a collateral soaking, so be it. Personally, I'd be just as happy if it died off sooner rather than later, removing my obligation to mow it again. As it is, I extract an old hand-mower from deep within the bowels of the garage and set to work before departing on the next leg of my journey. The hand-mower does a lousy job but it seems a fitting compromise; the lawn isn't so vigorous as to call down upon it the full wrath of the Rover.

I'm heading to Menindee Lakes, out where the recalcitrant Darling is forced to behave like a real river. It's a Sunday in early December as I drop my son at his cricket match and head out along the Barton Highway to Yass, then west and south down the Hume. As I pass the turn-off to Boorowa, the country is still looking luxuriant, retaining vestiges of spring green. But once I've crossed Conroy's Gap—at 650 metres, it's all that's left this far north of the Snowies and the Brindabellas—the fields start taking on the golds, the browns, the bleached bone and brittle whites of an Australian summer. I've crossed an invisible line: north of here rain has brought some relief, but to the south the worst drought in 100 years shows no intention of relinquishing its grip. There have been floods in Tamworth in northern New South Wales, creating a bonanza for cotton farmers along the Namoi, but little of that water will reach the Darling and none will reach the Murray. The rain could the harbinger of better times or merely a repeat of last year, when heavy rain and localised flooding in the north only accentuated the pain in the south.

The day is warming rapidly. Overhead the sky is radiant blue, but out towards the horizon the sun is already bleaching it to white. The song of the tyres moves from a low hum to a high whine and back again as the surface of the highway changes. The planet spins on, the horizons widen and the farmers of the western slopes hold their collective breath. On the car radio, the ABC reports on a Sydney man accused of murdering his elderly neighbour for watering his roses. The perpetrator incorrectly believed his victim was breaching water restrictions. It's being described as an extreme case of 'water rage'. At Coolac the traffic slows, funnelled into one lane by roadworks. St Peter's Catholic Church, a solid red-brick edifice built in 1925, has been auctioned; a bright blue sticker plastered across the real estate agent's sign declares it sold. Across the highway, a Beyond Blue billboard advises those experiencing depression to seek help.

Five miles from Gundagai I stop at the Dog on the Tuckerbox, and immediately regret it. A massive service station has sprung up, surrounded by a sun-blasted car park. As well as petrol, the 24-hours-a-day, seven-days-a-week complex boasts a Tuckerbox Express Restaurant, a KFC, a Subway and a Guru coffee shop. Evidence of these fine establishments is being blown around the vast parking lot by small willy-willies generated by the morning heat. Further along there's another fast-food palace, Buffalo Bill's, more temptation to lure the pilgrim from the path to the shrine—the shrine depicting in bronze the remarkable sight of a dog sitting on a box. 'A TRIBUTE TO OUR PIONEERS' declares the inevitable plaque. The memorial was erected in 1932, at the peak of the Great Depression. It cost £65 and, principally funded by public donations, was built in defiance of the local council, which refused to chip in. 'We have no money for roads, therefore we have none for monuments', harrumphed Councillor L Ryan, one of a long line of politicians to mistakenly believe that providing services trumps symbolism. Prime Minister Joe Lyons travelled from Canberra to unveil it, no doubt welcoming any respite from mounting economic disaster. Now, as financial crisis again sweeps the world, perhaps Kevin 'New Deal' Rudd will be inspired to travel the land unveiling more and more bronzes to our pioneering spirit: a plaque-led recovery.

The monument is, of course, a sanitised fraud, as is the song it's based upon. The original had it that the dog *shat in* the tuckerbox 5 miles from Gundagai, a yarn every bullocky would have known and loved: Henry Lawson's 'Loaded Dog' with attitude. But I guess it would have been difficult to convince Joe Lyons to put his name on a monument commemorating a dog taking a crap. Pioneering spirit can only be taken so far. So instead we've immortalised a dog for sitting on a box. I watch on as bewildered Asian tourists take their obligatory snapshots while they ponder the mundanity of Australia and just how mind-numbingly boring pioneering life must have been.

I drink a nondescript coffee at the nondescript cafe next to the legendary hound, but any chance of conversation is drowned out by a constant stream of bush ballads blaring from loudspeakers. So instead I read the laminated press clippings detailing Councillor Ryan's heresies, as ditty after ditty screeches out across the countryside: 'Hey True Blue' and 'Pub with No Beer' and 'Kookaburra Sits in the Old Gum Tree', followed by 'Click Go the Shears' and 'Red Back on the Toilet Seat', and, inevitably, 'There's a track winding back to an old-fashioned shack along the road to Gundagai ...' But there's no more track and there's no more shack. There's a four-lane highway, a car park the size of a football field and a reinforced-concrete fast-food emporium. I drain the coffee and hit the wallaby track.

It's not long before I leave the soulless expanse of the Hume, turning west onto the two-lane Sturt Highway towards Wagga Wagga. A farmer, taking a stand against the unholy tide that has taken St Peter's, has erected a banner: 'Wise Men Still Seek Jesus'. A short time later I'm hit by the locust plague. It comes from nowhere. The grasshoppers float hypnotically before the car, swirling with the lightness of grey snowflakes, yet they smack into the windscreen with the conviction of kamikazes, leaving their clear-and-yellow intestines splattered across the glass. The wipers soon do little more than smear them into a translucent goo, and I stop to wash away the carnage as hoppers alight amiably upon my head. Cars crawl pass with windscreens that look as if they've been pelted with eggs. A black-clad motorcyclist rolls by holding one gauntleted hand out to shield his visor: one of the four horsemen on a slow day. Across the way, set back from the road, is a vineyard. I can only imagine the damage being wrought there by the insects. Who'd be a farmer?

Past the garrison town of Wagga Wagga the land flattens, the road straightens and the car picks up speed. Off to the right, a thin line of trees, advancing then retreating then advancing once more, marks the passage of the Murrumbidgee. I pass the turn-off to Lockhart, 'The Veranda Town', and then Narrandera, 'Town of Trees'. The speed limit climbs to 110 kilometres per hour but it's observed as a minimum by the increasingly sparse traffic. Five hundred kilometres into my journey I reach Hay and cross another invisible line, beyond which travellers feel compelled to embrace the fellowship of the road, raising a lazy finger or two from the steering wheel to acknowledge passing cars. The temperature is well into the mid-thirties and I've taken to wearing a damp towel round my neck. With no air conditioning I've got the windows down, but even at speed the air is oven dry. Yet out here there is irrigation, with massive farm levees suggesting cotton or rice. I pull the car over and clamber up a 3-metre-high embankment, but the field beyond is barren. There's not enough water in the system; there are minimal allocations again this year.

The road leaves the river and heads out across the Hay Plain. The earth here is all but devoid of contours—not a gully, not a mound— given texture only by the stubble of its low grass. The land is so flat, the day so hot, that the horizon goes beyond shimmering and begins to dissolve altogether, with the bleached white-blue of the sky washing away the lip of world. Way off in the distance, oncoming trucks appear like supertankers afloat off the coast of reality. Telegraph poles march out into this sea like the piles of a long-gone pier. Off my starboard bow a lighthouse appears, its beacon flashing seductively, tempting me to drift away into the world of perception. It reveals itself as just another truck, carrying new cars east from the factories of Adelaide, its beacon the sun bouncing off the windscreens of its cargo. Another phantom, this time the arches of a bridge, floats into view off to port, hovering in the pale blue that should be land but has instead melded into sky. It's a huge linear spray irrigator, run aground on the shoals of nowhere, awaiting the next high tide to lift it back into production.

And then, 800 kilometres and ten hours after leaving Canberra, the country abruptly changes. Grapevines start appearing, their green leaves impossibly iridescent after the flat tans and palette-knife duns of the plain. The neatly ordered rows of citrus plantations stabilise the horizon, and olive groves spring from soil that moments before supported nothing more than mallee scrub. Before I know it, I cross the Murray River and arrive in Mildura. Paddle steamers and houseboats laze serenely on the broad water and birds are everywhere, frolicking loudly in the gathering cool of the sunset hours.

Later, I drink cold beers and eat pizza at a sidewalk cafe, watching as Mildura's citizens promenade in the relative cool of evening. I eavesdrop on the conversation of the twenty-somethings drinking champagne and debating the limits of statutory law at the next table. Maybe it's the beers, or perhaps it's fatigue or the clear summer light, but my vision has been polished with a filmic quality and I wonder if this is just a more-elaborate mirage. If I had travelled up the Murray from South Australia or down it from Albury, Mildura would appear in no way unusual, just part of the lush continuum of irrigated vineyards and well-watered parks that tracks the river. But drive to it from the north or from the south across the desiccated plains and you can see Mildura for what it is: an oasis out on the edge of the desert, totally dependent on the regulated flows of the Murray.

Lake Menindee is empty and has been for seven years by the time I walk out onto its sun-blasted wastes. 'Forty-six thousand acres of nothing', a local had warned me. The grey silt of the lakebed has been cracked open, leaving deep fissures. A few long-dead trees, their branches bleached skeleton white, offer no shade and less consolation. What little grass remains crunches underfoot: black, dead and tinder dry. I can detect no living thing. There are no clouds again today, and the black-powder bed is soaking up the sun as it once did water. I can't stay out here walking in the belly of the lake for long—the temperature is approaching 40 degrees and the heat is driving me back towards the shoreline. I feel vulnerable in this alien landscape, exposed, as if I should be wearing a spacesuit. I scan the distance, trying to comprehend the horizon-bending expanse, but it defies me. I retreat back across the moonscape to the white-sand beach and its gum trees, incongruously green. And up above me, perched on the lake's northern lip like some surrealist joke or a set left over from a Mad Max movie, spreads a long line of beach houses. This is Menindee's 'Sunset Strip', the weekenders built by the people of Broken Hill, 100 kilometres to the west. They sit above the beach on Lakeview Drive, respectable two-storey brick homes sharing the view with shacks knocked together from fibro and corrugated iron scavenged from mine sites and railway yards, all looking out across miles and miles of nothing. Barbecues sit on empty decks awaiting the return of water and, with the water, their owners. Seven years they've sat there empty, with their steel stairways extending down below the decks to the white-sand beach and the grey nothingness beyond. The houses have names like Casa Bella Vista and Possum's Palace. They speak of better times, of children building castles in the sand, of bikinis and board shorts, sausages and beer, sailboats and summer romances. I close my eyes and for a second I can hear the laughter, I can smell the barbecues and coconut oil. Bondi in the outback. And music, somewhere I can hear music. But it's not my imagination. Somewhere up along the Sunset Strip, someone has cranked up their stereo in defiance of the never-ending drought, and the notes cascade down into the lakebed.

I find Anthony Tolhurst sitting in the shade of his front veranda, taking a breather from working on his house. The music flows out of his house undiluted and crystal clear; on this windless day there is no competing noise. Lake Menindee is acoustically pure. Anthony himself is rake thin and wears no shirt, with the ginger of his long beard almost totally gone to grey. He pulls up a chair for me, offers a late morning beer. After my walk on the lakebed, it tastes like an epiphany. A latter-day hermit, he nevertheless welcomes the chance for company. He tells me he came here eighteen months ago, long after the lake emptied. Others may long for the water to return, but not him.

'I like the desert, I like the look of it. The way I see it, if the lake stays empty there are no boats, no jet skis and no people. 'Cos once that lake gets full again, there's lots of people who still have houses here', he says. He explains that the Sunset Strip is not abandoned; rather, it hibernates beneath the broiling sun. The electricity and town water are still connected. Of about 100 houses, Anthony reckons roughly a dozen are up for sale; another twenty or so have full-time residents. The rest simply wait. He's seen his neighbour on one side once in a year and a half, the one on the other side twice. 'That's plenty', he says. The permanent residents like to keep to themselves, occasionally congregating in small groups for a drink or a game of golf on the scrub course across the road.

'I escaped the city', explains Anthony. 'I sat there one night thinking, "I live in a city but I don't go to the city. I go to work, that's all I do. I don't go to movies. I don't go to bars". It cost me a fortune to live there. My long-service leave was coming up. So I thought, "Right". I sold my house, got in the car, headed off, wound up here.'

Anthony bought his house for $55 000 and has been fixing it up ever since. He says he came here to draw and paint, but for now he spends his days, and sometimes his nights, working on the house. He's poured a concrete slab and erected a galvanised-steel workshop on top. He's dedicated a corner beside the house to making home brew, and he's built an idiosyncratic barbecue, a robust construction with a hotplate and an oven, shaded by a canopy welded out of discarded car bonnets. 'The one on the top's from an XD Falcon, the one at back's off a CM Valiant', he states proudly.

Now approaching fifty, Anthony tells me he once ran his own graphic design business, drove a Jag and was hauling in the money, but a series of failed relationships left him disenchanted with the fast lane. 'I like my own company. I haven't been with a woman since 1993 and I'm quite happy to stay that way. I've been married twice and had a de facto in the middle and lost the house on every one of them. I've never been happier, ever, since I've been here. It's just the freedom and the quiet. You can go for a walk any time you like. The other night I couldn't sleep, so I worked through until 6.30 in the morning. I was on a roll.'

I leave Anthony to his solitude 1000 kilometres west of Sydney and head towards Menindee proper, 10 kilometres away, a third of the way round the circumference of the lake. At the junction where the Sunset Strip road joins the highway from Broken Hill, a faded sign warns: 'Water in this lake may be very cold. Wind squalls can produce dangerous waves'.

⌒

The Menindee Lakes are the most vilified bodies of water in the entire Murray-Darling. Cotton farmers in Queensland and citrus growers in South Australia may spit long-distance insults at each other, but they are united in condemnation of the lakes. 'The Menindee Lakes evaporates more water than the whole of Queensland uses', asserted Jake Berghofer by the weir in Cunnamulla. 'The whole system, it's designed to maximise evaporation. It's just fucking insane', argued Chad Prescott beside a St George cotton field. And they have a point. The vast shallow lakes baking in the western sun lose more than 2 metres a year to evaporation: more than 460 000 megalitres annually on average, twenty times that amount when the lakes are full. There's not a person in the whole basin who can't think of a better use for all that fresh water. If only it were that simple.

Menindee itself is a nondescript little town, 110 kilometres east of Broken Hill and 300 desert kilometres north of Mildura. Claiming to be the oldest town on the Darling, Menindee was the base camp for Burke and Wills' ill-considered push to the Gulf of Carpentaria, but its utilitarian buildings betray little of its age or heritage. Lacking a main street, it's an ad-hoc bundle of houses held together like onions in a string bag by the main Sydney–Perth railway line to the north and by the Darling River creeping past its eastern and southern borders, with the barren shores of Lake Menindee 2 kilometres to the west. The sign on the road from Broken Hill states Menindee's population as 980 and its elevation as 70 metres. A few years ago, when a ne'er-do-well hanged himself in the primary school playground, someone marked the population down to 979, plunging the town into a round of recriminations and finger pointing that lasted long after the council restored the sign. But, as always, it's the elevation that tells the real story. That's 70 metres of energy to get the water back into the river and down to the lower lakes 1800 river kilometres away. The elevation bears testimony to the flatness of the land, a uniformity that defies attempts to harness the waters of the river. For its entire length, from its arbitrary beginning at the junction of the Barwon and the Culgoa 1500 river kilometres to the north-west, until it reaches Menindee, there's no opportunity to dam the Darling. There are no valleys, no gorges in which to build a deepwater storage. The river is left to meander its way across the great western plain, with only small town weirs built within the river channel itself to impede its progress. That's on those days when it feels like progressing at all, rather than just lying there in a sun-soaked torpor, lacking the inclination to proceed further. But at Menindee, a unique opportunity exists for engineers to impose some discipline on the wayward Darling.

The Darling is an antipodean watercourse, slack and disrespectful of European notions of how proper rivers should behave. It basks in the sun, indolent and self-satisfied like a teenager after a late night, rising from its languor only occasionally to go on a flood-fuelled bender, spreading out across the land, careless of where it goes or what it touches. Even in its southern reaches, as it merges with the Murray, it threatens to break apart into an ill-defined delta. The Great Darling Anabranch departs the main river channel at Menindee and snakes its own way to the Murray, 270 kilometres to the south. Most days the river north of Menindee is a trickle at best, leaving townspeople and graziers in dread of stagnation and another bloom of toxic blue-green algae. In summer, only half the water that passes Bourke will make it to Menindee. The day I arrive, no water at all is flowing down from the north. The mighty Darling lies unmoving in its channel, the dirty gutter that Lawson described. Real rivers, big rivers like the Mississippi, the Amazon and the Yangtze, don't do this; they don't just stop in their tracks. Nor do they fluctuate much. The difference between a dry year and a flood year in the Amazon catchment is a 30 per cent variation in available water; on the Yangtze it's 90 per cent; on the White Nile, flowing through the deserts of Sudan and Egypt, it's 190 per cent. On the Darling it's 470 000 per cent. It's either flood or drought; a cliché maybe, but the truth nevertheless. And, with no possibility of European dams, it's left to the Menindee Lakes to harness the floods as best they can.

Back before 1950, floodwaters rolling down the Darling from Queensland would overflow into Menindee's ephemeral lakes, large shallow depressions in the desert lying just to the west of the river's main channel. Then, as the floodwaters receded, the lakes would empty most of their water back into the river. The idea of using the lakes as a gravity-fed water storage was simple and cheap: let the water flow into the lakes at times of flood, but prevent it flowing back out until needed. And so between 1949 and 1968 a series of weirs, channels and levees were constructed. And it all worked well, or it did back in the days when there was enough water, back when it was an issue of regulation rather than preservation. Back then, no-one in Queensland complained about evaporation, and in South Australia, those who thought about it at all were grateful for the reliability that Menindee guaranteed.

Now, though, everyone wants Menindee to operate more efficiently, and inside the air-conditioned offices of State Water, local manager Mike Arandt takes me through the options being considered to cut evaporation. Mike's a practical, hands-on man, unburdened by degrees and theories and the political intrigues of faraway capitals. He came here in 1976 for three months and ended up staying, working his way up to be the top man. In many ways, the lakes are his life's work. But aged fifty-five, retirement beckons; he and his wife have bought a farm near her family in Canowindra. His impending retirement has liberated him from many of the constraints of his employment; unlike so many bureaucrats, he's willing to offer an opinion. And so he brings out maps detailing each of the six options being developed. It's a presentation he's made before: to state and federal politicians, to irrigators and environmentalists, to journalists and anybody else willing to travel the distance and take the time. But it's not a presentation grown stale by repetition; he may be leaving soon, but Mike cares about what will happen out here.

'It doesn't matter which solution you come up with, whether it's number one or number six, you're going to have some people who support it and a lot of powerful people who oppose it. It's very hard to get a balance', he says. An hour and a half later, with my brain swimming in gigalitres and my eyes blurred by diagrams and flow charts, Mike takes me out in his four-wheel drive to show me firsthand what the maps represent. Little matter that it's long past five o'clock and knock-off time: this is his domain and he wants the faraway people of the coast to understand what is happening here.

He takes me out to the main weir, a dam-like structure that raises the level of the river 12 metres and is the key to the Menindee Lakes system. Lacking a valley in which to store the water, engineers have fabricated one, building levees that stretch more than 30 kilometres from the weir up the eastern side of the Darling. To the west, the water flows into four relatively small natural lakes. The whole thing, flooded river and flooded lakes together, has been given the name Lake Wetherell. We climb up upon the weir's superstructure and stand maybe 15 metres above the water. Lake Wetherell stretches off into the distance, the drowned skeletons of red gums marking the original river course. Cockatoos screech overhead and pelicans float serenely on the surface. Over to one side, a flat-bottomed boat lies moored at a small wharf: a cruise boat available for tours on request. Lake Wetherell works well enough but it doesn't hold all that much water, at least not enough to stretch a good Queensland flood out over three or four years of drought. It's almost full now, but just a year ago, water levels in the system were down to 1.5 per cent. So at times of flood, water can also be diverted into three much larger natural lakes: Parmamaroo, Menindee and Cawndilla. Parmamaroo can be filled by water diverted from the main weir and can empty directly back into the river. After Wetherell it's the first filled and the last emptied. If Parmamaroo fills, an interconnecting channel can then fill Menindee. It too can empty directly back into the river if water levels are high enough. The third of the lakes, Cawndilla, fills from Lake Menindee and either empties back into it or diverts water down a series of shallow creeks into the Great Darling Anabranch. Neither Menindee nor Cawndilla have had water in them for years.

Mike takes me out to see Parmamaroo. It too was empty a year ago, before northern floods sent enough water down the river to fill it. This caused a flurry of accusations from South Australians distressed over the state of the Murray, the lower lakes and the Coorong. Why was water being fed into the evaporation pan of Parmamaroo when it could have been sent down the river? Mike tells me that a sizeable amount of water did go down the river once Parmamaroo was filled because a decision was taken not to put any water into Menindee. And in the year since, water from Parmamaroo has helped supply irrigators in the southern Darling and on into the South Australian Murray. Now the lake is emptying once more as water is sent downriver before it evaporates away. Its surface, shining in the evening light, has receded a couple of hundred of metres from the lake's edge, leaving behind a swathe of green grass. A small mob of kangaroos looks up as we pass before returning to the rich grazing.

But there's a problem here. Before the weirs and channels and regulators, floodwaters would rush into the lakes and rush back out again. Now, the water, heavy with silt, rushes in but is drip-fed back out again. The result is that the mouths of the lakes have silted up. 'Imagine a water tank', explains Mike. 'On the side of the tank you have a tap. If it's at the bottom you can get all the water out. But the silt is stopping us getting all the water back out. It's like the water tap is halfway up the tank—everything below it you can't get at.' That was one of the reasons no water was let into Lake Menindee during the previous summer's minor flood—much of it would have remained inaccessible. I look out over what remains of Parmamaroo, at all that water sacrificed to the sun, and I share some of the indignation of the South Australians.

Menindee is the largest of the lakes in terms of surface area, but Cawndilla is deeper and has more capacity. Its greater depth means less evaporation, and its water is less salty and of better quality. Not surprisingly, four of the six proposals to improve the efficiency of the lakes encompass a greater role for Cawndilla. The problem is that the lake fills from Lake Menindee, not from the Darling itself. Elaborate and expensive plans have been drawn up to divert water around Lake Menindee and directly into Cawndilla. An even greater hurdle is getting the water back out into the river again. The proposed solution involves a new channel being carved from Cawndilla's south-eastern corner back to the Darling. But the channel would bisect the oldest national park in western New South Wales. Worse, much worse, it would plough directly through Aboriginal burial sites. Not just one or two graves, but hundreds of them, maybe thousands—no-one knows for sure. And so it is that the premier engineering solution would be the most damaging to the environment and the most offensive to local Aboriginals.

'It's the best way to make the lakes work as a system, but the impacts on the Aboriginal people are huge', observes Mike. 'And we put them under enormous pressure by saying to them, "We don't want you to make up your mind, but what do you think?" And if push comes to shove and they say, "No, we don't want it", the irrigators are going to say, "Fucking black bastards wouldn't give it to us". But if they say, "Yes, you can have it", their own people are going to go crook at them and say, "Why did you give in to them and why have you wrecked our national park?"'

For a moment we look out over Parmamaroo in silence, verdant grass lapping at the retreating water, as the sun edges closer to the horizon. We're up on a high embankment at the lake edge, built up over centuries by waves and wind, similar to the ridge above Lake Menindee where the Sunset Strip has been built. Finally, Mike breaks the silence. 'You know, where we're standing, these dunes, they're covered in middens. They're here, all around us. This view, for thousands of years the Aborigines would have looked out on exactly this same view. Makes you think.'

⌣

That night in the pub I chat with Daryl, who's wandered across the road from his motel for a beer. It's a Monday night and Bill, the barman at the town's other hotel, has shut up early and come over for a beer himself. We talk with Bungy behind the bar. The three men have all come from Broken Hill and all profess Labor roots. But they have become more and more disillusioned with the passing years, willing to offer up their votes to any politician decent enough to pay their isolated town a visit. They hear snatches of the criticisms of Menindee coming from far away. All agree it's ill-informed. They don't think much of Canberra, yet most of their disdain is reserved for Sydney.

'N.S.W. You know what it stands for, don't you?' asks Daryl. 'Newcastle, Sydney, Wollongong. They don't give a stuff for anything west of the sandstone ridge.'

'Sandstone ridge?'

'Yeah. What do you call it? The Blue Mountains?'

The next morning Mike takes me out in his truck again, as if to make sure I don't miss a thing. He takes me along the levee to a quiet stretch of river, wide and bountiful. It's like the Murray down by Mildura but without the people and the intrusion of civilisation. I know it's not natural, but it looks wonderful.

'There's a beaut place in there, just around that corner, where mobile phones don't work. I spend a bit of time in there. Don't catch many fish, but it's quiet', he chuckles. We go down to another much smaller weir where the water leaves the lakes system and heads off downstream. There's a steady flow going over the weir: water to supply downstream irrigators, water for the ungrateful South Australians. Mike tells me that every few years the weir needs to be restored. 'See those rocks there, those circles beside the weir? Every time we rebuild the weir, they come back.'

'Fish traps', I say.

'Good. You know about them. I've never seen them make them, but when we come out to check the weir, they've often been moved about.'

Beside the river the trees are strong and healthy, fed by the near-constant flow coming from the lakes, a contrast with the thirst-racked red gums up near Tilpa and Louth. But back from the river the land is dry, the sparse grass interspaced with areas of bare earth. I ask Mike how bad the drought has been.

'Well they say it's the worst in a century, and maybe it is in terms of rainfall, but the country around here is still in better shape than the drought of the early eighties. Back then, there was no ground cover left at all. All this soil was simply blowing away. It was the rabbits, they ate everything, then the topsoil just blew away.' The rabbits are now under control, thanks to the calicivirus. A few emerge during the winter months, but once the hot weather comes down upon the land, they die off once again. So the land has had some respite, and so have the small native marsupials that compete with the rabbits. But rabbits are notoriously adaptable, and no-one expects the calicivirus to remain effective forever.

Talk of rabbits leads me to ask about carp, the rabbits of the river. 'They came here in 1976', says Mike. 'We had a flood, a really big one. We were out working on the main weir. We had the gates open, letting the water through, and you could see them swimming up against the flow. Nobody knew what they were.' But Mike says now the native fish are making a comeback, and there are less carp in the river. One theory has it that the carp can't handle the hot still waterholes of the drought, another that the Murray cod have learnt to eat carp fingerlings. I climb back into the truck, pondering the land's resilience.

Mike takes me out to a sandy flat sitting high above a bend in the river. The ground is hot and flat and, to my untrained eye, nondescript. My guide points out the remains of campfires where the sand has been blown away, the charcoal staining its whiteness. I pick up a piece that looks a week or two old.

'These fires are hundreds, maybe thousands of years old', Mike explains, bending down. He picks up a fragment of shell and then a stone tool.

'Is this a burial ground as well?'

'No', says Mike. 'Not here. That's not for me to show you.'

He takes me out onto Lake Menindee, unaware I've already ventured here. He pulls the truck up beside a long-dead tree. For the third day running there are no clouds, and the grey-black lakebed is layered in blankets of heat. The tree is a scar tree, bearing the evidence of not just one but two scars where, long ago, Aboriginals stripped away the bark to make bowls or small shields. I know what Mike is trying to tell me, but it all seems rather remote, belonging to another age, back when Burke and Wills passed through.

And then something catches my eye, lying bleached white against the dark claypan. It's a mussel shell. Not some tiny pipi but a big thing, the size of my closed fist. It immediately brings to mind Fred Hooper's talk of feasting on mussels as a kid up at Weilmoringle. I bend down and pick it up. The two shells are still held together, just as they were in life. It has weight, substance.

'How old is this?' I ask.

'Seven or eight years. Last time the lake was full', says Mike.

And for some reason, the reality of the mussel does what Mike's careful explanation of the burial grounds of Cawndilla has failed to do, sparking what the ancient middens and campfires didn't ignite: some inkling of just what the bounty of these lakes, and that of the whole river, must have meant to the Aboriginal people. And still does. I hold the mussel in my hand and can see the gathering people, not the ragtag groups of purposeless nomads that inhabited my schoolbooks, but hundreds, perhaps thousands of people coming together as the lakes fill with water from far away. I see the fishing and the feasts, the re-establishment of old bonds and the making of fresh ones; corroborees and trade, children splashing in the water, love and grief under the river red gums. I see an ancient and resilient culture replenishing itself with the waters of the Darling, a population large enough to create the vast burial grounds of Cawndilla. I hold the mussel and contemplate the vacuum of Lake Menindee.

⌣

Weeks before, up in Brewarrina on the Barwon River, up above where it becomes the Darling, I'd seen fish traps that dated back between 16 000 and 18 000 years. That's why I recognised the ones Mike showed me near the weir. At Brewarrina I'd talked to 73-year-old Les Darcy. He'd grown up in a tin shanty beside the fish traps. In his childhood the Aboriginals had still lived there, as they had in antiquity. At the beginning of the twentieth century the blackfellas were still required by law to be back across the river by nightfall and the sergeant of police would come with a stockwhip to enforce the ordinance. Les had spoken of the river with reverence. 'Biami created it. He was our creator. Him and his two sons and his spirit dogs come along and created the river and all the other parts of eastern Australia. And when he came to Brewarrina he saw that the people were starving and there was no water. So he taught them how to build the fish traps and then he taught them how to call the rain. And then they called the rain, and when the water subsided there were thousands of fish in the river. That was our creator.'

Les had said that the river used to run clear and clean, that he and his family used to drink from it, that for much of his youth he didn't like rainwater because it didn't taste right. 'There was plenty of fish and crayfish and mussels. There was always something you could get from the river … I don't know why they built a weir because we had a natural weir which had never [been] known to be dry in any man's time. They put a weir on top of our fish traps. Not beside. On top. Some of our fish traps they dug up with bulldozers and dynamite. And we've got something there that we can safely say is … one of the oldest man-made structures on earth. Perhaps the oldest. And what do they do? They blow it up with bloody dynamite.'

At the age of nineteen I'd hitchhiked round Britain and had set out for Stonehenge, my imagination stoked by everything from *The Lord of the Rings* to Arthurian legend. I'd been shocked by what I found: a major highway running hard up alongside the ancient stone circle, the asphalt sucking out the magic of the site. Well, my belated apologies to the English. At least their road runs beside Stonehenge, not right through the middle of it. Nor did they resort to dynamite. And their stone circle is merely 4000 or so years old, nothing compared to the stone circles of Brewarrina's fish traps that date back almost 18 000 years. And in 1974 we built a concrete weir smack-bang in the middle of them. Whatever were we thinking?

'It was estimated there were 5000 people here at one time at the fish traps. That was in white man's time, so how many would there have been in their own time?' Les had said. I'd asked him to come down to the river with me, to show me the fish traps for himself, to point out where his family's tin hut had stood and describe how it had once been. But he couldn't do it. He couldn't bring himself to look upon it, not with me there to witness his distress.

'It's embarrassing to go to the river', Les had said. 'It's not the river anymore. It's completely changed. We'd have weeds going out 10 or 15 feet into the river. And we'd have crayfish and shrimps and all of these little fish swimming in it … All that's gone with the chemicals they've used for the cotton now. We've got nothing in the river now. They've ruined our river—the cotton people. And that was only in the sixties that they started to ruin our river. The seventies, when they built the weir, they really got into it. It was the ruination of Aboriginal life as we lived here for thousands of years.'

Many of the river towns of western New South Wales and Victoria have sizeable Aboriginal populations. In most, the blanket racism of the past has been replaced with a cautious willingness to judge individuals rather than communities, but problem towns remain. Bourke is one, Wilcannia another, and Brewarrina is one of the worst—drugs, alcohol, abuse.

'There's nothing here', Les had said. 'There's no work, there's no employment. There's nothing for the kids except drugs and boredom. Talk about kids getting into problems; if we had water and did things the way it should be done, it would help our young people. It's sad, it is. The river was our lives, and when they take the river from us, what have [we] got?'

In place of Les, a member of a younger, angrier generation, Brad Steadman, had taken me down to the river. The stones of the fish traps, hauled in from far away somewhere between 160 and 180 centuries ago, still lay in the riverbed below the concrete weir. Brad had explained how people used to herd the fish into smaller and smaller circles until they could simply bend down and pluck them out of the stream. But the traps, in continuous use for millennia, were now seldom if ever used. He explained how the stone circles, without anyone to tend them, were gradually being buried by the silt left in the lee of the weir. He told me that the industrialisation of the river had merely continued what guns and missions and influenza had begun: separating the traditional people from the river.

And yet nothing on the Darling is simple. Not the arcane systems of irrigation licences and water markets that dictate its unnatural flow, and certainly not the relationship between the river and its traditional guardians. Irrigation, more than drought or climate change, has taken the natural river away. But where irrigation is strong, there's money and there's work.

Up in St George, I'd visited Ron Waters and his dad, 80-year-old Kevin. Kevin never got much of an education; in 1932 his dad had removed him from the mission where he'd lived in New South Wales and taken him to Queensland so he wouldn't be stolen away by the authorities. 'They would have taken me for sure. I got fair skin. They always took the fair-skinned kids', Kevin had told me without apparent rancour. Ron was different. A member of the Kamilaroi Land Council, he'd been educated and city-trained; he knew how to look out for himself and for his people. Ten years ago, as the cotton wealth flowed through St George, he'd realised with mounting anger that his people weren't getting their share. Local farmers were employing backpackers or bringing in workers from elsewhere. But when he approached local growers he found it wasn't racism that was stopping the Murri from getting work; it was their own unreliability. Typically, they'd show up for work for a few days and then go walkabout. So Ron had set up an employment service. The growers would call him, state how many workers they needed and for how long, and Ron would guarantee the workforce. Early each morning he'd drive round town and see who wanted to work. So the growers didn't get the same workers every day, but they didn't mind as long as the work was done. Indeed, most were only too happy to help the local Aboriginal community. Everyone was happy, and gradually the flow of money started washing away some of the entrenched disadvantages of the Aboriginal community, rinsing out some of the town's prejudice with it. 'We had 120 or 130 workers at one stage, before they started using genetically modified cotton that doesn't need as many workers. The Murri here have lived off the cotton industry. Cotton and grapes', Ron had said.

A similar scheme was started in Bourke. Local people worked in the vineyards out at Back o' Bourke Fruits. But that's gone now. Irrigation at Bourke is collapsing, and so are employment opportunities for local Blacks. It's even worse at Brewarrina and Wilcannia. The natural river has gone, and now the unnatural one is going too. Yet at St George there's work. And at Menindee, there's employment maintaining the lakes and caring for the vineyards that tap their water. The Indigenous people of Menindee hold strong familial bonds with those in Wilcannia; all agree that the social problems of Menindee are small compared to those of its northern neighbour, and that the availability of work is part of the reason why.

⌣⟶

I ask Mike Arandt if it would be possible to meet any of the local Menindee elders. He explains there are only five, and three of them are crook; arranging a meeting at short notice may be difficult. But 71-year-old Harold Bates welcomes me into his house not far from the banks of the Darling. It's a proud home, with family photos adorning the walls of the neat lounge room. Harold bought the house with his own money, earned from years spent working for the water authorities. He's a shy man, uncomfortable talking to an outsider, especially one armed with a notebook, a voice recorder and a back-catalogue of ignorance. It's obvious the only reason he's talking to me at all is because of his regard for Mike. And it's only the encouragement of Mike, sitting across the room, that gets him to talk at all. The trust and friendship between the two men is evident as they recall rugby championships from decades past, children raised and the passing of the years.

Eventually I touch on the sensitive issue of Lake Cawndilla and the proposal to carve a channel between it and the Darling, a channel that would desecrate an untold number of Aboriginal burial sites. But Harold surprises me. He's retired now but for many years he worked alongside Mike for State Water. Like Mike and a handful of others, he understands the intricacies of the lakes system, where the water flows and where it doesn't. Like Mike, he understands that Cawndilla and its channel would deliver the best quality water with the least evaporation. But Harold knows he must oppose the Cawndilla proposal. He's an elder with a responsibility to his people, and his people don't want the graves destroyed.

'They're all against it, so I'll have to be against it too, yeah. My wife reckons it'll bugger up the national park putting a big river through there. And it will be a big one.'

And so I sit in Harold Bates' shady living room and contemplate the affection and respect Mike and Harold hold for each other, and the irony of their respective positions: Harold appreciates the technical logic behind using Cawndilla and privately doesn't have many qualms about building the channel, but he's destined to oppose it in solidarity with his people; Mike privately opposes Cawndilla because he understands the immense impact it would have on the Indigenous people, but is required to remain officially neutral on the proposal. On the Darling, nothing is as simple as it first appears.

Weeks later, I sit in the front bar of a pub in a river town. Across the road, beyond the immaculate lawn of a park, the river lies brown and sleeping in the bottom of its channel. It's early evening, the sun backlighting the trees by the river, a perfect end to a perfect day. After a long afternoon of driving, the pub offers respite: a cheap room upstairs with a bathroom and toilet at the end of the corridor; a counter meal downstairs with a couple of cold beers. The bar is quiet, with just four or five regulars gathered in the corner for a round or two. They are old men mainly, in their sixties and seventies, neatly dressed in their shorts, sandals and long socks; the youngest would be in his fifties. I'm too tired to introduce myself, but there is something comforting in their banter, something familiar and reassuring as they crack old jokes and laugh companionably. So I have half an ear open as I write my notes and listen to the interviews I've conducted that morning. And, as the conversation touches on Aboriginals, I flick on my voice recorder to capture it verbatim, as amid the laughter it swings through a pattern I've heard before.

'They used to say, "Sambo was a lazy coon, he wouldn't work in the afternoon". But you're not allowed to say it now, 'cos it's racist.'

'The bloody way things are, you'd think the Christ fellow was a coon.'

'I'd tell you what, you'd be right.'

'We'd be bloody millionaires by now.'

And like a river rounding a bend, the conversation moves on to other topics. The comments pass without challenge, without debate, unremarked; among the old men they are in no way contentious. They're just old jokes, lazy prejudices restated for no particular reason. I think of Menindee and Brewarrina and half a dozen other places.

Outside the pub the sun is setting, and as the artificial lights take over, the interior of the bar loses its comfort. I leave the rest of the beer and go for a walk beside the river.

5

THE INLAND SEA

Lake Hume near Albury, 6 per cent full

Iblame Matthew Flinders. I thought of accusing Napoleon Bona-
parte, but that would be drawing too long a bow. Flinders
achieved many impressive things: he discovered Bass Strait and
established that Tasmania was an island; he circumnavigated Australia
and charted much of its coastline; and he helped coin and popularise

the word 'Australia'. But in early April 1802, he blundered. Sailing east in command of the *Investigator*, Flinders was charting the coast of what is now South Australia when he encountered French explorer Nicolas Baudin, the captain of *Géographe*. The two rivals were said to have exchanged information and pleasantries in the spirit of science. But at a time when Napoleon's navy had already fought Nelson at the Battle of the Nile, and was yet to fight him at Trafalgar, perhaps the two navigators could be forgiven if they were distracted from their cartography. For both men, as they drew their charts, missed something noteworthy: the mouth of the Murray River, down where it enters the Southern Ocean. The place is called Encounter Bay, in honour of the two navigators who missed its most significant landmark. Or perhaps I do the rivals an injustice; perhaps the Murray wasn't flowing into the sea that autumn. Whatever the reason for Flinders' oversight, it had a profound impact on Australian inland exploration. For when European explorers began crossing the Blue Mountains a decade later, they found more and more rivers carrying more and more water westward. And as Flinders had established that these rivers did not flow into the ocean, the explorers concluded, not unreasonably, that they must end instead in some mighty inland sea, somewhere off in the heart of the continent. So I blame Flinders for sending these explorers scampering off into the hinterland, searching for the ends of the rivers and ignoring their origins. More to the point, it was Flinders who unwittingly helped plant in the imagination of a new country the dream of an inland sea and a desert made green, a dream that endures two centuries later.

Elsewhere in the world, explorers became famous for heading up rivers, not down them. As a schoolboy, I learnt all about Stanley and Livingstone seeking the fountainhead of the Nile, and Lewis and Clarke heading up the Missouri, but what Australian student learns who discovered the source of our mightiest river? We may have heard that a Polish count named Strzelecki climbed the country's highest mountain and named it Kosciuszko, but who was it that discovered the wellspring of the Murray not 50 kilometres distant? It was Thomas Scott Townsend, in 1845. Australia's second highest mountain, Kosciuszko's neighbour, is named after him. I've walked the slopes of

Mount Townsend, and skied them in winter, without once wondering how it got its name. But Townsend didn't so much discover the Murray's source as survey it, and its authenticity was decided not by him but by politicians. How very Australian.

It was well known by the time of Townsend's expedition that the Murray rose in the Australian Alps, but no-one had been particularly fussed about which particular mountain brook should be declared the actual source. Some rivers, like the Nile and the Mississippi, spill ready-made from great lakes; others flow complete from the ends of glaciers; a rare few burst forth spectacularly from the bases of mountains, like the Sorgue at the Fontaine-de-Vaucluse in France. But the origins of most rivers are less definitive. They begin where a multitude of creeks join together, and who is to say which is the principal and which is the tributary? Such was the case with the Murray. The source only became an issue as pressure mounted for Victoria to become a separate colony. At first the Murrumbidgee was suggested as the border, but the mother colony soon vetoed this, not wanting its southern rival to emerge with the lion's share of riparian wealth. So the Murray became the border—not split down the middle, as is usually the case with waterways, but with the river entirely within New South Wales and with Victoria starting at the high water mark on the southern bank. Yet with the border decided in principal, no-one was entirely sure where the river began, and so Townsend was dispatched to survey the various creeks, streams, brooks and springs that fed it. Even then, it wasn't Townsend who declared the one true source, but an Act of British Parliament. In 1851, the Act of Separation defined the boundary between New South Wales and the new colony of Victoria as the Murray River: 'on the north and northeast by a straight line drawn from Cape Howe to the nearest source of the Murray River'. So the border defined the source, not the other way round. It wasn't until 1869 that another surveyor, Alexander Black, confirmed a spring high in the mountains as the river's most easterly origin and therefore its official source.

I've decided to see the source of the Murray for myself. It seems only proper if I'm writing about the river, if along its length I find it degraded and corrupt, that I see it where it first begins, up where it's natural and free. By now summer is well and truly entrenched. Christmas has come and gone in a flurry of wrapping paper and brandied pudding, and a new year, 2009, has snuck up on us under cover of fireworks and champagne. I study my maps, buy my provisions, plan my assault. Locating the source on a map is simple enough: it's where the squiggly line marking the border between New South Wales and Victoria joins the straight border line that stretches down to the coast. But how to actually get there?

I read that on his Nile expedition, Livingstone led a band consisting of freed slaves, Comoros islanders, twelve sepoys and two servants; I take along my 10-year-old son, Cameron, who is growing bored with his school holidays and his hoard of Christmas presents. As we drive south towards the mountains, he bounces with anticipation around the front seat of the rented four-wheel drive: we're going bushwalking, camping in the mountains; it'll be his first time sleeping in a tent. At Jindabyne, overlooking a lake formed by the Snowy Mountains Scheme, a towering Soviet-style statue of Strzelecki points off into the main range, a visionary gaze chiselled into his face. But there is no statue, no plaque, to commemorate Townsend. Little is known of the surveyor apart from his work, although one report has it that he went insane in later years. Perhaps he had a premonition of what was in store for his river.

From Jindabyne, Cam and I drive south along the Barry Way—all mountain meadows and shiny gums. Small white clouds adorn the sky, so perfect in their proportions and position that they appear confected. The road turns to gravel and we plunge into the alpine forest of the Kosciuszko National Park. But the forest doesn't last long. We climb over a ridge and reach the Wallace Craigie lookout, which takes in a vista spreading south over the valley of the Snowy River and into Victoria. The view goes on forever, but it's a sparse, almost barren panorama. In 2003, the same bushfires that incinerated more than 500 homes on the western rim of Canberra came through this valley.

The bush is still recovering. Some trees are dead; others still display the furry signs of epicormic regeneration, shoots growing directly from the tree trunks. In the distance, the Victorian Alps fold away in diminishing shades of blue, but on the nearer ridges the gaps of earth between the trees are all too evident. The fire here was so extreme it generated a giant tornado, its gargantuan throat emitting a roar heard more than 5 kilometres away. The smoke was driven so high into the atmosphere that it formed a thunderhead that in turn rained lightning strikes down into the tinder forest, sparking new blazes to join the firestorm. Six years later, the evidence of the apocalypse is everywhere. Cam, who clearly remembers the day Canberra burned, is momentarily silenced. But only momentarily. Then he's scrambling around in the dust in delight, retrieving the large flakes of fool's gold he's spotted gleaming in the sun. Another four-wheel drive pulls up, and an elderly man joins me at the lookout.

'Looks like it's on the way back', I say.

'Mate, it's only going to burn again', he states with grim certainty.

It's a warm day, almost hot, in the high twenties. We drop down into the valley, where the Snowy appears to have garnered at least some of its promised environmental flow. For an unsealed road it's an easy drive, and a constant procession of four-wheel drives is running in both directions: city dwellers off to find some of the freedom promised in the brochures. Down by the river, semipermanent camps have been set up that boast tents the size of small houses. What better place to spend a week or two to escape the heat? Yet as we cross the border and start climbing into the unburnt Victorian high country, the temperature begins to fall. It's down to 14 degrees as we cross the 1400-metre-high Wombargo Range. The sun is still well above the horizon, but mist begins to drift out of the high forest of mountain gums.

'Look, Dad, look!' exclaims Cam. A wallaby bounds from the bush and across the road. I brake hard. 'It's a brush-tail. A brush-tail! They're endangered, Dad. Wait 'til I tell Mum. Wow!'

We drive down into a valley and make camp at Native Dog Flat. Cam bounds around, exploring, while I swear my way through erecting the tent. That night, as my son sleeps, brumbies thunder past, snorting and braying, their hooves shaking the earth. I cautiously pop my head out, hoping for a glimpse, but the bush is impenetrable in its blackness. There is no light. And yet, when I hold out my hand to confirm my blindness, I find I can see its outline after all. The night is moonless, but above me the stars are blazing. I climb to my feet, the dew cold on my bare skin, and behold the heavens. The Milky Way is sprayed from one horizon to the other, the dark shapes of the nebulae clearly delineated against the creamy wash of the galaxy. The sky is alive, the earth is dead. I hold my breath, ignoring the growing numbness in my feet. Somewhere to the south, a microscopic piece of dust hits the atmosphere and streaks white across the sky. I decide I should wake Cam, share with him the celestial dome, but back inside the tent he breathes the unconfused sleep of the young. I let him sleep; tomorrow will be a big day.

The next morning, after striking camp, my plans go bung. The track I'd identified to carry us within striking distance of the Murray's headwaters proves to be nigh on impassable. Perhaps a heavy-duty four-wheel drive might be able to take it on, one equipped with a winch, a chainsaw and off-road tyres, but not my lightweight hire car. Not with my son in it. So I hastily revisit the maps. There is a way, but it means a much longer walk. We wind our way up the main road to the top of the ridge, crossing the Great Dividing Range, and enter the Murray catchment before turning onto the Cowombat Track. Cam delights in the name: 'What's a cowombat, Dad?' The maps show the track winding all the way to the Murray, but vehicle access is restricted. Sure enough, before long we reach a gate, and set off on foot into the Cobberas Wilderness. The day is shaping as a hot one; we carry plenty of water in our daypacks.

Through the bush we walk, side-by-side, the track easy and wide. Around us the forest is healthy, but not dense. Fire has been through here as well, although it's recovered far better than back on the Snowy. Cam is astounded by the mounds of brumby shit that adorn the track every few hundred metres, especially their size.

'Why do they all do it at the one spot, Dad?'

'They don't mate. That's just from one horse.'

'Cor. How big are they?'

There are flowers everywhere: delicate five-petalled pink flowers with a yellow stamen, royal bluebells, mauve daisies with yellow-button centres, bushes covered in white heather-like flowers, alpine buttercups. Cam insists I take photos of them all to show his younger sister, Elena. While I click away he rushes ahead, wanting to see what's round the next corner, over the next slope. I tell him to pace himself, that we have a long way to go. Even so, after an hour's walking we're in such good spirits that I feel overly cautious enforcing a five-minute break and making Cam sit down and drink some water. Not long after we start again, a lyrebird sails across the track ahead of us, wings outstretched, long tail trailing behind. We rush to keep it in sight as it lands in the bush. Half an hour later a stallion pushes out of the forest and stands for a moment on the track. He's coal black, imperious, the sheen of his coat rippling with light. A horse from a storybook. He snorts his contempt and canters unhurriedly into the bush, moving away up a gully. I steal a glance at my son. He is standing with his mouth open, his eyes round, transfixed. As we continue on our way, I ask him if he thinks brumbies should be allowed in the national park.

'Sure. Why not?'

'Well, they're not native animals.'

'They're not?'

'No.'

He thinks for a while. This is a boy who loves his native animals, but who has just encountered a horse as magnificent as the imagination.

'Well, I guess they could let them stay, if they could just get them to stop crapping everywhere!'

The morning wears on. The terrain becomes more difficult. We climb up steep mountain ridges, down the other side. The hourly breaks become half-hourly breaks and no longer need to be enforced. The sun climbs higher in the sky and the forest offers less shade. The temperature is climbing into the thirties. Cam's feet are starting to hurt: he has his mother's flat feet and he's not used to walking so far and for so long. Coming down a steep hill, he slips on gravel and scrapes his knee. Blood comes and tears can't be far behind. As I bandage the knee I remember he's only ten years old. What am I doing dragging him along on such a route-march? Already we've been walking for more than three and a half hours. I consider turning around. 'No,

Dad', he says adamantly. 'We need to get there for your book.' Some 10-year-old. I check the maps, taking my compass readings from the surrounding hills. It can't be much further but the track is steep now, with few flat areas to provide easy walking. We're also getting low on water. We push on.

After four and a half hours of walking through the forest, we emerge from the trees into a wide alpine meadow, another storybook image. The grass is green and soft, dotted with wild flowers. Cam, who has grown quiet and truculent, is immediately cheered. 'Is this it, Dad? Is this the Murray?' And it is. Down the slope of Cowombat Flat, a thin line cuts the grass. We walk down together, past an explosion of white flowers. We behold the mighty Murray in its infancy, a mountain brook half a metre wide, flowing icy clear over a bed of stones. Someone has erected small signs on either side of it: 'NSW' and 'VIC'. We take photos of each of us straddling the border, one foot in either state. Then I take Cam over to a large gum, standing alone in the meadow on the News South Wales side, and we collapse in the shade. I take Cam's shoes off, give his feet a rub, put bandaids where blisters are threatening to emerge. He's so tired that he can barely eat lunch. To make matters worse, he cracks one of his remaining baby teeth and has a hard time eating his apple, despite his hunger. The actual source of the Murray is not far, just a kilometre or so—a spring pushing through the native grass a few hundred metres off a walking path. But it's already half past two in the afternoon. I'm not dragging Cam any further, nor am I about to leave him alone in the wilderness while I go by myself, no matter how gentle the breeze, no matter how idyllic the pasture. Instead, I let him doze in the shade while I return to the creek. In places it narrows to just a few centimetres as it passes over rocks and through Lilliputian gorges. It's as clear as gin. Small fish dart around a small pool; I don't know what they are but they're not carp. I cup my hands and drink the pristine water: it's cold, ice cold, and somehow sweet. It's such an important river, the Murray— so important to our history, our economy, our sense of nationhood. Yet here it is just begun, like Cam, giving few clues of how it will grow and change on its long journey to the sea. For the moment it plays and gurgles down tiny waterfalls, an insubstantial thing that I could momentarily dam with nothing more than a well-placed boot. Up here in the mountains, up above the farms and the towns and the politics, it's pure and unpolluted. I drink as much as I can and fill our bottles.

The meadow itself is a wondrous thing, somehow appropriate here, as if nature recognises that this river requires something special, that just another bush-clad gully wouldn't suffice. I wonder what quirk has left it open when all around is forest. We have seen no other such clearings all day. Off to the north rise the towering heights of the Pilot Wilderness, covered in the silvery fur of trees dead from fire and dieback. Further west, the valley of the Murray carves down and out of sight, mountain peaks folding blue into the distance. The far side of the valley appears covered in poplars, but they're gum trees, trunks bristling with growth as they slowly recover from fire. To the south-east, the khaki forest climbs dense up to the ridge of Mount Cobberas. In winter, the pasture would be blanketed in snow. I promise myself that next time I'm on a plane to Melbourne I'll keep a weather eye out for it, not far south of Thredbo, in the shadow of Mount Pilot.

With some food and a rest, Cam has regained his verve, but I wonder how long he will last. Before leaving Cowombat Flat we investigate the remains of an old plane, a DC-3, that crashed back in the 1930s. The crumpled aluminium lies untarnished, shiny in the sun. Cam asks what happened. I tell him I don't know, but it's not hard to imagine: a pilot in trouble, losing height, with nothing below save forests and mountains, ridges and cliffs, when, like an answer to a prayer, he spies the open field through the clouds; one last desperate throw of the dice. Cam and I regard the mangled remains of the superstructure and I leave the end of the story unspoken.

The temperature is in the mid-thirties and we drink more and more of the clear Murray water as we retrace our steps. Rests are no longer timed; we walk until Cam reluctantly tells me he needs to stop. The sight of an occasional brumby pack cheers us momentarily, but for the most part we walk in silence. We reach a creek, where we again refill our bottles and soak our shirts in the cold water. By now my legs are also beginning to ache, and I break out the emergency chocolate. We don't talk anymore; there's not much to say. I lead and

Cam trudges behind. But he's cracking hardy, and every time I turn and wait, he gives me a big smile. My calves are hurting, my thighs are aching, and my hips are contributing their own harmony of protest. Finally, the temperature begins to ease ever so slightly as the sun heads down. And still there is a long way to go. I'm carrying warm clothing and torches just in case, but we're not equipped for a night in the open. We'll just have to keep walking until we get there. Then Cam starts laughing. 'You know, Dad, this is going to be a family legend, isn't it?' We reach the car just past eight o'clock, as the sun is touching the horizon. We didn't reach the source, but we got to the Murray, up high above the world, up where there's nobody to spoil it. We haven't seen another soul in ten hours. And now that we're back in the car and Cam can rest, I figure it's been worth it.

⌒

The next day we call in at the Murray 1 Power Station. Giant white pipes slice straight down the mountains' western flank, feeding the station's turbines. Power lines, buzzing with energy, stretch away down the valley. The day is blazing hot: 42 degrees. Thank goodness we aren't walking today. Cam emerged this morning from the tent to stretch in the sun like a cat, complaining of stiffness. Good for him. I can hardly put one foot in front of the other. Overnight my hips have frozen up, and I waddle around with the gait of an 80-year-old on an elective surgery list. At least today is a driving day; the hire car is an automatic and I don't have to lift my leg to operate the clutch. The power station is nestled not far from where the Murray, by now a real river, emerges from the mountains and starts its long journey westward across the ever-flattening plain. But the water that spins the generators inside Murray 1 is not from the river; it comes through the mountains, diverted by the Snowy Mountains Scheme.

Inside the cool of the visitors' centre, the woman behind the counter gleefully tells us we've just missed the guided tour. 'Thank God for that', I reply, happy to keep any walking to a minimum. Instead I take in the dioramas, explain the relief map to Cam, and grab the first seat I can to watch the film explaining the scheme. I swear it's the same one that was playing here when my parents brought my sister, brother and I through here as kids. 'Nineteen sixty-three was a royal year …' intones the narrator over pictures of a glazed-eyed monarch; 'showing keen interest, Her Majesty proceeded to the tunnel face …' Even without the royal seal of approval, the Snowy Mountains Scheme stands as Australia's largest, most ambitious engineering scheme. But as I gaze out the window at the water flowing from under the power station and off towards the Murray, I wonder whether it should stand alone, or whether it would be better viewed as part of a continuum, just one step in a 200-year effort to make real the dream of an inland sea.

It was Charles Sturt who discovered Flinders' oversight and established there was no inland sea. After an inconclusive expedition in 1828, following the Macquarie, the Bogan and the Castlereagh deep into what we now call the Northern Basin, and discovering the Darling along the way, Sturt headed down the Murrumbidgee in 1830 with the stated purpose of determining once and for all where the western rivers ended. He followed the main channel until it merged into an even more significant river, which he named the Murray, not realising it was the lower reaches of a river that Hume and Hovell had found in 1824 and had modestly christened the Hume. Sturt followed the Murray, discovered its confluence with the Darling, and journeyed on to within sight of the sea at Encounter Bay. It was a massive achievement, but Sturt must have felt disappointment; he'd established that all the major western rivers flowed into the ocean and not an inland sea. And yet in his book *Two Expeditions into the Interior of Southern Australia during the Years 1828, 1829, 1830 and 1831*, Sturt was enthusiastic about the land he saw to the west as his men rowed him across Lake Alexandrina:

I would repeat, as my view of it was, my eye never fell on a country of more promising aspect, or of more favourable posi-tion, than that which occupies the space between the lake and the ranges of St. Vincent's Gulf, and, continuing northerly from Mount Barker, stretches away, without any visible boundary.

In England, Edward Wakefield, who had been casting about for a location to establish a colony of free settlers, read Sturt's account and decided that South Australia fitted the bill. The Englishman didn't procrastinate: Adelaide was founded at the end of 1836. Sturt himself moved to the new colony shortly afterwards, herding sheep from New South Wales to help build the economy of the fledgling settlement. But Sturt appears to have been haunted by the elusive mirage of an inland sea. In 1844 he set out full of hope into the very centre of the continent, but could not penetrate the Simpson Desert. Still not satisfied, he mounted another expedition into the central deserts, but succumbed to scurvy, went temporarily blind, and was lucky to survive.

The South Australian colonists quickly realised the Murray held the key to prosperity. In 1850 the government offered a £4000 prize to the first two iron-hulled boats to reach the Darling. Francis Cadell built a paddle steamer in Sydney and in 1853 brought it through the sandbars at the mouth of the Murray into the lower lakes. From there he travelled 2400 kilometres up the river and claimed the prize. By the end of the decade the Murray had been navigated as far upriver as Albury, the Murrumbidgee as far as Gundagai, and the Darling as far as Bourke. The new colony was on a winner: as the Murray itself drained the inland rivers of eastern Australia, the South Australians were perfectly positioned to drain the growing pastoral wealth of New South Wales and Victoria. The first railway line in Australia, initially a horse-drawn affair, was completed as early as 1854 between the river port of Goolwa on Lake Alexandrina and Port Elliot, 13 kilometres away on the coast. But the geographical advantage enjoyed by the South Australians gradually eroded as the railways in the eastern states extended out to the rivers, connecting Melbourne with Echuca in 1864 and Sydney to Bourke in 1885. Nevertheless, by this time the Murray had firmly established itself in the minds of Adelaide's citizens as the link to the east and to the future. Nowadays, the city's residents are again taking an interest in the river: not in its promise, but in its decline.

I put my contemplations aside and hobble around, looking for Cam. I find him over by one of the displays—not the one on Snowy Hydro's commitment to the environment, nor the one espousing the benefits of renewable energy, but the one on water. 'The scheme's most vital role is to provide a reliable supply of water to the Murray and Murrumbidgee rivers for irrigation and river management', it states.

⌒

Back in Canberra, summer proper hits like a sledgehammer. A high-pressure weather system parks itself over the Tasman Sea, pumping hot dry winds from the deserts down across Australia's south-east. The temperature hits 40 degrees and stays thereabouts for a fortnight. The lawn dies, bleached and brittle, and the mowers are silenced for another year. The leaves on a silver birch in the front yard go from green to brown inside five days and start to drop months early, and still the heat mounts. A wattle in the backyard is dead, a tree that cheered us with its defiant blooming during the depths of Canberra winters. The earth under our house becomes so dry, the water table retreating to such depths, that the soil contracts and new cracks start appearing in the walls of our 45-year-old home. The thick brick walls that have kept us cool through previous summers now soak up the heat and beam it back at us during sleepless nights. And we're not having the worst of it, not by a long shot. At Renmark in the South Australian Riverland, the temperature soars above 48 degrees. Even farms with water struggle to withstand the heat as fruit cooks on the trees and vines wither under the assault. Finally, reluctantly, the blocking high starts to drift towards New Zealand. But it leaves its worst for last, sucking along in its wake winds oven-baked from the west, turning Victoria into a fan-forced kiln. The forecasters can see it coming but can do nothing save issue warnings. The garden state burns. Some 173 die, towns destroyed, communities shattered. Communities little known outside Victoria—Marysville, Kinglake, Strathewen—become bywords for grief and devastation. Not since Ash Wednesday in 1983, during the last great drought, has Victoria suffered such a catastrophe. And in the north, up in Queensland, floods have come, as if to taunt those marooned in the parched-brown and burnt-black land of the south.

I wonder what I'm doing here in Canberra, why I'm not back

down the coast where there's a sea breeze and a cool ocean to plunge into. More to the point, I wonder what Canberra is doing here. What tin-pot sadist decided to maroon us here, stranded from the beach, every Australian's birthright? Every January we Canberrans drive the Kings Highway to Bateman's Bay in our tens of thousands and spread out along 200 kilometres of beaches. So why isn't the city there permanently?

The heat is too oppressive to work at home, too hot to travel, so I retreat to the National Library's air-conditioned main reading room. The books tell me a new capital was made necessary by the pre-Federation squabble between Sydney and Melbourne; neither could tolerate the other becoming the seat of government. That much I knew, but why inland? To protect it from naval bombardment: that was the accepted wisdom of the founding fathers, back when they debated the constitution in the last decade of the nineteenth century, back before the advent of the aeroplane rendered such considerations obsolete. Yet it's an intriguing thought: Australians of the day felt exposed sitting along the coastline, with no cities to retreat to in the hinterland. But I wonder if that's all there was to it. The debates over the constitution occurred throughout the 1890s, the same era when the *Bulletin* and other publications were mythologising the bush, when the countryside was being advanced as the spiritual cradle of a new country. Is it possible that an inland location was seen as alluring, and not just sensible?

The constitution simply said that the 'seat of Government of the Commonwealth … shall be in the State of New South Wales, and be distant not less than one hundred miles from Sydney', and left the actual location to the parliament. By the time federal politicians started touring possible sites in 1902, all of the short-listed locations were inland. In 1903 a ballot in the House of Representatives chose Tumut, while a vote in the Senate chose Bombala. The impasse was finally broken when Canberra-Yass was settled upon. It was a typical political compromise; who can say what motivated individual parliamentarians to settle on Canberra-Yass? But there was consensus that the capital should be sited in the hinterland. And so I am left sweltering in Canberra either because our founding fathers feared bombardment, or because they felt a visceral need to centre the new country in the bush.

Finally, the heat of summer starts to retreat. The kids have returned to school and the weather is benevolent enough for me to depart for the rivers once more. But I drive into a country licking its wounds. The day I leave, the sky is still hazy with smoke from the fires lingering in Victoria. The ABC reports that an algal bloom has closed Canberra's Lake Burley Griffin to swimming and boating for the rest of the summer. As I head down the Hume, the country beyond Conroy's Gap looks battered. I wonder how the rivers are faring.

The spread of the railways heralded the beginning of the end for the natural river. Once Melbourne was connected by rail to the Murray at Echuca (1864) and Wodonga (1873), and Sydney was linked to Bourke (1885) on the Darling, to Gundagai (1886) and Wagga Wagga (1878) on the Murrumbidgee, and to Albury (1881) on the Murray, Victoria and New South Wales saw less and less need to buttress the rivers for navigation. Instead, the eastern colonies were becoming increasingly interested in irrigation, initially to drought-proof grazing land, then to grow crops, and finally to help populate rural areas. But the rivers were proving fickle. The Murray ran dry during the Federation drought of the 1890s, and again in 1903. Even in a good year, the river would run strongly in winter and spring but become shallow and sluggish during summer, halting not just riverboat traffic, but depriving irrigators of water when they needed it most. Eventually, in 1914, the Commonwealth, New South Wales, Victoria and South Australia signed the Murray River Agreement, which still forms the basis for managing the Murray today.

Under the agreement, a huge storage, the Hume Dam, was constructed just upstream of Albury to capture the spring snow melt and the winter rains that flowed down the upper Murray from the Alps. Once the dam was completed in 1931, the Murray never ran dry again, as the water in Lake Hume was rationed out over summer. At the other end of the river, barrages were built across the channels where Lake Alexandrina met the Coorong, ensuring the lower lakes remained a freshwater environment year-round. With Hume Dam guaranteeing the water supply at one end, and the barrages acting like a one-way valve at the other, the river had become a sealed system. In-between the two end points, a series of twenty-six weirs and locks had been planned, stretching from lock one at Blanchetown in South Australia, above the lakes, all the way to Echuca in Victoria, with another nine weirs and locks intended for the Murrumbidgee. The water backing up from one weir would stretch all the way to the next weir, guaranteeing navigation. In the end, eleven weirs were built as far upstream as Mildura on the New South Wales–Victoria border, a distance of 878 kilometres from Goolwa on the lower lakes. But that was far enough; after failing to find the fabled inland sea, Australians had built it instead. For that is what the Murray is now, from the lakes to Mildura: one long thin stepped lake, or a canal, or an inland sea, where the water level is near constant, year in, year out; where pleasure craft can motor from one end to another; where prosperous communities grow produce in desert soil made fertile by its waters.

The system was developed further after World War II as Lake Hume struggled to supply the increasing demand for irrigation water. Hume Dam was expanded, the Snowy Mountains Scheme began feeding water through the mountains, and in 1979, Dartmouth Dam, a deep high-altitude dam in the Victorian Alps, was completed. The Menindee Lakes were engineered to store and regulate water coming down the Darling, allowing irrigation to begin on the lower Darling. But no matter how much new water was put into the system, expanding irrigation was taking it back out. The engineers toiled mightily, but struggled to keep pace with demand.

⌒

One hundred and eighty metres beneath the top of Dartmouth Dam, the solid granite walls are cool and damp, a blessed relief from the noonday sun. Peter Liepkalns, the man who operates the dam, stops the car at the end of the steeply inclined tunnel, and we climb out into the man-made cavern. We're in the valve chamber, deep in the living rock beneath the dam wall. Towering metal machines, controlling the four valves, climb towards the ceiling, condensation gathered on their grey-painted surfaces. There's a church-like quiet in the still air, and I find myself almost whispering as my voice echoes off the stone and concrete walls. Not so Peter. He's brought me down here to explain the operation of the dam and the dramatic impact of the long dry. Beneath us, through a series of steel grates, lies the diversion channel. It was built to reroute water around the construction site as the dam was being erected in the 1970s. After completion, the channel was only to be used rarely, when the higher outlets were taken out of service for maintenance. Now the diversion channel is the only outlet that can be used.

'It was never designed to be used all the time, because under the initial modelling this dam was only ever going to be below 80 per cent full once every ten years', Peter says. 'When the dam was built, no-one ever envisaged climate change or that we'd ever get this low.'

We climb back in the car and Peter expertly reverses the half-kilometre up the steep incline back to the surface. We drive towards the top of the dam wall, eyes blinking as we readjust to the light. The view out over the reservoir is sobering. The remaining water lies low in the lake. Roads from the construction days have resurfaced, and the pale white ghosts of long-submerged trees stand high and dry. The high-water inlet, a three-storey concrete tower, sits beached 20 metres or so back from the water's edge. Water should be pouring through the top of the tower to feed the hydro power station below the dam wall, but the power station has been mothballed since Christmas 2006. This year the dam is down to 21 per cent of capacity. Last year it dropped to 17.5 per cent; the year before to 11.9 per cent. Below 7 per cent, not even the diversion channel can drain the remaining water. Being here, seeing the lack of water, reading the concern in the eyes of the man who manages the dam, makes stark the reality in a way statistics on a computer screen cannot. For the past three years, the entire Murray system has been running on empty.

For Dartmouth Dam is the Murray's storage of last resort: a deepwater reservoir kept high in the mountains, the ultimate guarantor that the inland sea will never run dry. The water level behind the weirs and locks has remained unchanged since the 1930s, and it's

Dartmouth's role to keep it there. Back before Dartmouth Dam was finished, the Hume Dam near Albury carried that responsibility. These days, the two work in tandem. They hold roughly the same amount of water, but the Hume is emptied first; its shallower basin is more susceptible to evaporation and its larger catchment refills the dam more than twice as quickly. But my guide is worried. 'I was listening to the ABC the other day and they were saying we're heading for another El Niño. Dry weather. But it's not wet now, it's dry. You can't get drier than dry.'

No-one knows more about Dartmouth Dam than Peter Liepkalns. He first came here as a kid, the dam the last of seventeen construction sites where his father, Jan, had worked. He recounts the time he wagged school to visit his favourite fishing hole, of how his life was saved by the sharp eyes of a passing workman who spotted his bicycle hidden in the undergrowth, how the detonation was stopped just in time, of how furious and how relieved Jan had been. Jan, now eighty-four, came to Australia as a refugee from Latvia in 1947, together with his German wife, a survivor of the Dresden firestorms. By the time the family got to Dartmouth, Peter's mother had had enough of constantly moving, and told her husband it was time to stay put. So Jan moved from construction to operations, and worked his way up to the position his son now holds. Peter went off to university, then returned for what he thought was eight weeks work back in 1980. Now, at the age of fifty-four, he's not even considering retirement. 'No, no, no. No way. Bit of a family heirloom this. Can't let anyone else look after it. When I retire, Jan will have to come back', he says, poker-faced.

Peter's an unpretentious sort of bloke, no airs or graces, no bureaucratic babble, his face sunburnt and open under a cloud of untidy hair, the blond curls almost totally grey. He reminds me a bit of Mike Arandt, out at Menindee: someone who knows a lot, someone who cares a lot. Someone who cared so much that when the bushfires swept through in 2003, he drove through the flames in the middle of the night to manually adjust the valves on the dam as the power station was taken offline. It's not something he wants to talk about. It's more important to talk of what the flames did to the catchment.

'The whole lot burnt', he says. 'There wasn't a bloody green leaf left, apart from the green belt, for 100 kilometres. Then it got burnt ten days later. Ninety-five percent of our catchment was burnt. Decimated. You go up the back of the bush now and there is so much undergrowth and stuff we don't get the run-off. We haven't had the rain, but the little bit we have had, the run-off is minimal. The other day we had fifty mils of rain, Thursday three weeks ago, and the lake didn't even respond fifty mils. So where does it go? A bit of hot weather, a bit of evaporation, but if it didn't fall in the lake, it didn't get in there.'

Dartmouth Dam tells the story of the drought and its impact on the Murray. As inflows dried up in the early years of this decade and the Hume was all but drained, the authorities fell back on Dartmouth's strategic reserves, essentially borrowing against the future, trusting that the drought would break as it always had done in the past. But by the summer of 2006–07 the dam levels had fallen to 11.9 per cent, perilously close to the 7 per cent cut-off. Irrigation allocations along major stretches of the river were cut to nothing as authorities contemplated a worst-case scenario: river towns running out of drinking water, the inland sea running dry. Elsewhere, that scenario was dangerously close to reality. In Goulburn, a city of 20 000 people, Level 5 restrictions were imposed and plans hurriedly drawn up to pipe in water. That could not be allowed to happen to the Murray, so the farmers were cut off and Dartmouth's strategic reserve was gradually rebuilt. By last year it had risen to 17.5 per cent and should bottom out at its present 21 per cent this summer. But Peter Liepkalns takes little comfort from the slowly rising levels; they have not come from increased inflows but from starving farmers downstream of water. He says that nothing prepared him and the dam's managers for times like this.

'It's a one-off. We've got nothing to compare it with in 100 years of records and modelling. This doesn't fit in with any pattern they've ever recorded, so what is it? Hopefully it's not permanent, but all the experts are saying it is. In 1982–83 the dam fell to 9 per cent but immediately bounced back. That year mirrored the three other major droughts we've had. Back in 1898 and in 1907 and whenever the third one was, each of the major droughts we had was followed by a very wet number of years … and we were back in high-level water in no time. But we haven't been going up consistently since 2000. Our inflows have been at record lows for the last seven years. It goes up a bit and you draw it down. It comes up a bit and you draw it down some more. We've been on a downward trend since then. We haven't got the water to transfer and we are here as the dam of last resort.'

Driving downstream from Dartmouth Dam, following the Mitta Mitta River as it emerges from the mountains, the eucalypt bush gives way to a landscape of bucolic serenity, a diorama imported from far-off England. The grass is green and lush, the sky blue and cloud-speckled, the river clear and strong. Cows graze in the shade of willows and oaks; a family of ducks paddles near a dozing angler. The river winds its way through a postcard of well-defined banks, past small beaches of sand and stone, under the branches of imported trees. It's the sort of river to win the approval of the *Oxford Dictionary*. For 20 or so kilometres the illusion holds. Little matter that the water is being trickled out from Dartmouth's concrete diversion tunnel; for a few precious moments it's possible to pretend that nothing is wrong, that the river is unchanging.

But as the land flattens and the Mitta Mitta moves down towards the Murray, reality returns. In the larger perspective of the expanding brown valley, the river and its thin green ribbon is reduced to a trickle. My map tells me there is a lake here, Lake Hume, and so do the roadside signs. But the turn-offs are to picnic spots in the middle of nowhere, for there is no lake to be seen, just the far-off creek. Only the carcasses of drowned trees prove the lake has ever reached this far up the valley. Another signs informs me that Tallangatta is 'The Town that Moved in the 1950s' but you wonder why it bothered. The old town was flooded when the Hume Dam was expanded and the new Tallangatta was built on a rise overlooking the larger lake. But from Tallangatta, it's not possible to see its waters, not even in the distance. I drive down to the town's concrete boat ramp, past the sign banning boats with accommodation or toilets from the lake, past the sign warning of blue-green algae. But the boat ramp ends in a field parched brown, and what's left of the river lies a kilometre away across the sunburnt plain. The locals, though, are philosophical. The lake is always drawn down by the end of summer. They're growing used to this elongated tide, where the lake does its best to refill over winter and spring and all but empties over summer and autumn. This year it's down to 6 per cent capacity.

I drive across the bridge in search of water, climbing over the peninsula separating the Mitta Mitta and Murray arms of the lake. Off to my left, the density of once-submerged trees increases and, after topping a rise, I eventually see what's left of Lake Hume. Not much. I stop at Bellbridge on the opposite shore from Hume Dam, where the lake is at its deepest. The town's boat ramp gets closer to the water than at Tallangatta, but it still bottoms out 50 metres short. Behind me, half a kilometre away, the houses sit expectantly on a ridge. The sight elicits a sense of deja vu. I chase down its elusive tendrils, tracing them back: Lake Menindee's Sunset Strip.

Canberra's inland location put it beyond naval bombardment; the Snowy Mountains Scheme's location put it beyond aerial bombardment. At least, that was one of the justifications advanced by the Chifley government when it proposed the scheme. Opposition leader Bob Menzies opposed it, and Melbourne and Sydney had their own reservations: Victoria would lose control of the Snowy River; New South Wales had its own plans to divert the Snowy into the Murrumbidgee. And so Chifley invoked the Commonwealth's defence powers to give his legislation a constitutional base. He argued that Australia's coal-fired power stations were vulnerable to enemy attack, while the hydro stations buried deep within the mountains would give Australia a strategic reserve. That was soon forgotten once the scheme got underway and Menzies, upon becoming prime minister, became one of its strongest advocates. It was no longer some defensive stratagem; it was nation building. Ultimately it was more about water than electricity, and the country's subliminal desire to green the desert and populate the heartland. I recall the signs that Cam and I had read together at the Murray 1 Power Station: 'The scheme's most vital role is to provide a reliable source of water to the Murray and Murrumbidgee rivers for irrigation and river management'; and

'During drought, the Snowy Mountains Scheme can supply more than a third of the water flowing into the Murray River, as well as water to the Murrumbidgee River'.

I stand by the edge of Lake Hume and look along the shoreline. A few remaining posts from a 50-year-old fence line stand clear of the water. The bridge stands high and ungainly above the lake, its stanchions spindly, their foundations exposed by the shallow waters. Six per cent full. How full would it be without the Snowy Mountains Scheme? How full would South Australia's inland sea be? And what about Albury-Wodonga, just 5 kilometres away below the dam wall— where would it be?

I drive over Bellbridge's exaggerated bridge into New South Wales and wind my way round to the dam wall. The nearby Hume Resort has the feel of a ghost town, but the woman behind the counter at the cafe is upbeat. 'It's always like this mid-week, once the school holidays are over. But it's a long weekend this weekend. We'll be full.' I ask about blue-green algae, saying I'd heard the lake had been closed. 'No, not here. That's up in the Mitta Mitta arm. Everything here is fine for the moment.'

I sit by myself out on the cafe's deck and eat lunch, watching a group of council workers basking in the sun. Below me the swimming pool and waterslide are closed, awaiting the weekend throng. I understand why South Australia in particular was so keen to build the inland sea, and I kind of understand why Canberra was sited inland, and it would be un-Australian not to understand the Snowy Mountains Scheme. But Albury-Wodonga? In 1972 Labor won power federally for the first time since Chifley, and Gough Whitlam championed the establishment of inland growth centres at Albury-Wodonga and Bathurst-Orange. Albury-Wodonga was the principal site; Bathurst-Orange was a sop to New South Wales. At the time, Albury-Wodonga and its surrounds boasted a population of around 50 000; Whitlam declared it would have 300 000 by the turn of the century, another Canberra. Perhaps Whitlam wanted to build something to rival the capital, which itself had been a stuttering town of around 50 000 just fifteen years earlier, back before Robert Menzies finally decided to fill Burley Griffin's lake and start moving government departments up from Melbourne. Upon retiring from politics, Menzies declared Canberra as one of his three great achievements, not something that his party of free enterprise mentions all that often. Perhaps Whitlam wanted to leave his own monument; his contemporaries nicknamed Albury–Wodonga 'Whitlambad'. The stated reason for bolstering Albury–Wodonga was to relieve the mounting populations in the sprawling conurbations of Sydney and Melbourne. But why inland? Why not Coffs Harbour, or Two-Fold Bay? Why the Murray?

I finish my lunch and continue on into Albury, past the suburban subdivision of Thurgoona, a product of Whitlam's grand project. The radio tells me that more than 200 workers have been sacked at a local car component manufacturer. The hinterland may be less vulnerable to naval bombardment, but there are no defences against the global financial crisis. Of all the inland towns of the Murray–Darling Basin, Canberra excepted, Albury–Wodonga has diversified away from an agrarian base, linking into the national and global economies. But its history and its present are still linked to the river. In Albury's riverside park I find the Hovell Tree, where the explorer William Hovell carved his name and the date: November 17 1824. He and Hamilton Hume were the first white men to discover the Murray. The tree is still standing but the inscription is hidden away behind a barrier of plexiglass and folds of chicken wire. Nearby, the paddle steamer *Cumberoona* sits tied to a low wharf. Like most things of any age along the river, it's painted in heritage cream and dark green. A sign blocking the gangway says 'Boarding 15 minutes before departure. Tickets available on board'. But while the steamer itself looks in good condition, the noticeboard near the gangway is empty and growing derelict. A pleasant woman at the tourist information centre tells me the boat hasn't run for two and a half years. 'Not enough water', she says. 'Most days I could wade across the river there and it wouldn't come up to my knees.' So much for the 'mighty' Murray River. The ABC News tells me the South Australian Government is going to the High Court to challenge Victoria's limits on trading water interstate.

Gordon Craig, the man Whitlam appointed to build his vision, lives twenty minutes from town in a rural subdivision developed by the Albury-Wodonga Development Corporation, which he headed for a decade and a half. At seventy-nine, Gordon retains the guile that helped him pull off the sort of bureaucratic manoeuvre that would leave Sir Humphrey Appleby speechless with admiration. Whitlam lavished more than $200 million on the grand project in its first two years. But with a change of government, Malcolm Fraser cut expenditure to $5 million a year, barely enough to pay salaries, and recommenced Menzies' project, moving what government agencies remained in Melbourne to Canberra. But Craig had seen it coming. He wasn't about to relinquish his elevated status as a first-division officer, the equivalent of a departmental head in Canberra. Nor did he want to return to his former life as a civil engineer and town planner, to be frustrated by the limitations of local government. This was his one and only chance to achieve something of lasting value.

'For years, I had worked for towns like Shoalhaven, Murray Shire, like Young, for Wallerawang on the edges of Lithgow, trying to get those towns to develop into something worthwhile and real and attractive, and at last this was a chance to achieve it', he tells me over tea and scones. 'It didn't surprise me that the government withdrew its support, but by then we had spent about $200 million dollars on buying land and property. In other words, I had a bank. That's why I stayed. That's what I did in my first two years: I established a bank so I could work on Albury-Wodonga. So I didn't work for the government, I worked for myself. Because it was a project that I thought was worth doing and I had $200 million dollars to start it off with.'

Gordon Craig ran his personal fiefdom for sixteen years, skilfully playing the Commonwealth and state governments against each other, until in 1990 the three governments finally called a halt and closed down development activities, setting a deadline to sell off what remained of Craig's land bank. Craig retired, but he believes Albury-Wodonga, the second-largest inland city in Australia after Canberra, with a population of about 100 000 people, has achieved much of what he wanted: a self-reliant, diversified city. When I ask him why Albury-Wodonga and not some coastal location, he says that relieving the pressure from the major cities was the principal rationale for decentralisation, but there was also a widespread desire to stop a perceived decline in the bush. 'There was this concern that the capital cities were growing too big and that we shouldn't have the country towns getting smaller—they had been getting smaller ever since the motor car was invented and since bitumen roads came into existence … The idea of developing centres away from the capital cities had been debated for years. I can date it back to 1960 and probably before that.' Craig says Whitlam and his Minister for Urban and Regional Development, Tom Uren, were only different in that they were willing to act in a grand manner.

It was the Labor government that decided on Albury-Wodonga—other sites, like Bathurst-Orange, were only identified to pacify the states. Craig says Albury-Wodonga was chosen because it was far enough from Sydney and Melbourne to grow independently from them; it was on the major highway and railway link between the two, and it was already a centre of some size. And its location on an inter-state border helped justify the intervention of the Commonwealth. But there was one more reason: the river.

'This is the sort of place … people from Melbourne look at and say "We wished we lived up there"', says Craig. 'They come up here for their holidays in droves. We're littered with wineries and holiday resort cottages and the whole works. And the river is the focus of that. The wineries irrigate from the river; the weir is a major attraction. A lot of the rivers, the tributaries, come in from the snowfields and feed into the river. They're all part of the same mystique.'

Tracking back into Albury, I decide to take another look at the weir. Back in 1996, the structure moved 4 millimetres, which doesn't sound like much but was enough to put the local paper in a lather and the town in a panic. The lake was emptied and multimillion-dollar remedial work was undertaken. As I drive along, a brown heritage sign catches my eye: 'Bonegilla'. The name echoes. 'Bonegilla Migrant Hostel' repeats a second sign. I take the turn-off. The suburban street winds through a subdivision of brick veneer and roller doors before emerging among a small collection of corrugated-iron barracks, painted a neutral cream and lifted off the ground on brick pilings.

A larger sign tells me this is all that's left of the site of the Bonegilla Immigration Hostel. I drive around the deserted buildings until I see a new structure off to one side, with a silver four-wheel drive parked out front. Inside, I find John De Kruiff working behind a makeshift desk. John tells me I have stumbled across Australia's largest migrant intake camp, once known as 'Little Europe'. I'm the first person to visit all day, and John, a retired policeman, is all too happy to tell me about the place.

Bonegilla was built as an army camp during World War II. John says a lot of the military's heavy equipment was warehoused in Albury because of the different railway gauges in New South Wales and Victoria—storing it in Albury meant it could be dispatched to the designated port without the need to load and unload at the border. After the war, the empty barracks were perfect for a migrant intake camp. And so it was that hundreds of thousands of migrants had their first real taste of Australia not in Sydney or Melbourne, but at Albury-Wodonga. I look at the corrugated-iron barracks: stinking hot in summer, frigid in winter. Some people only stayed a couple of nights, others for weeks, many for months. Some stayed for ten years or more after finding work in the camp.

'At one stage there were over 6000 people living here. And there was also 1600 people living in tents ... the population was over 7500, so it was bigger than Wodonga was in those days. Three hundred and twenty thousand people passed through here between 1947 and 1971', John tells me. 'They had banks, they had churches, they had a school. They had a hospital. They had a picture theatre.' John's information centre is decked out with huge black and white photos on the walls. One shows a family eating a meal: the father in a suit coat and white collared shirt, the mother in a cardigan, the three children in knitted woollen jumpers. 'Dinner: Roast Mutton & Gravy' reads the caption. The next photo shows a wedding reception, men dressed in suits, women in their best dresses, seated at trestle tables with short white tablecloths. 'Lunch: Boiled Meat & White Sauce'.

'Sounds appetising', I say.

'Yeah', John laughs. 'In early 1952 the Italians rioted here. Over food. The thing I've heard more complaints about than anything else was mutton. Because this was a sheep area, the farmers around here had old bloody sheep, old boilers, that they couldn't give away, so they sold them to the federal government. They slaughtered them here, and people were eating mutton here three meals a day. And I've had Germans, mainly Germans, coming here and telling me that even fifty years later, if they smell mutton cooking they have to walk out of the room. What a lot of people did here, they'd go into Albury and buy a little kerosene stove and they'd either go hunting for rabbits or they'd go fishing in Hume Weir and catch fish and cook up the fish.'

'Really? Is the lake far from here?'

'Are you joking? It's just there.' And he points out the window to a levee less than 20 metres away. John tells me the levee wasn't built until the 1970s and that the lake was visible from the camp, just a three- or four-minute walk away. And so the migrants weren't just dumped in the middle of nowhere. Thanks to an abnormality in rail gauges, their first experience of the Australian hinterland was not of a parched and arid land, but of a heartland blessed with water; a camp sitting on the edge of a large and plentiful lake where generations learnt to swim and, according to John, nineteen drowned.

John takes me for a walk through the old dormitories and family rooms. In an old mess hall, he and his fellow volunteers have set up a small museum, complete with photos and memorabilia, that goes some way to communicating the Spartan camaraderie of the camp. He points out an enlarged photo on the wall of a beaming family, apparently taken at sea. At the front, a young blond-headed boy looks shyly at the camera. It's John, then known by his real name, Jan. Another Jan, come to build Australia. His family spent six weeks at Bonegilla in 1952, not long after the Italians had led the mutton uprising. John says he doesn't recall much of the camp, just a couple of fractured memories.

I ask him what happened to his parents. He tells me his father had owned a dairy farm in the Netherlands, and so he got work locally milking cows. His parents never left the district, living out their lives within 20 kilometres or so of Albury. They weren't alone. Many of the migrants never made it back to the coast. For not far away there was work for the asking: the Snowy Mountains Scheme. And so the families and single men of Little Europe moved into the construction towns of Cooma and Jindabyne, Cabramurra and Khancoban, and a half-dozen more little Europes. There, as they became Australians in the mountain camps and worksites, they helped Australia grow beyond a land of mutton eaters. There wasn't room for the ethnic ghettoes of big city immigration; the different nationalities were forced to work together. The migrants worked with their hands to build a new Australia, not realising they were really building a new country with their hearts.

And the influence is still there, stretching down the river, in the vineyards and the olive groves, in the cafes and restaurants, in Griffith and Mildura and Shepparton and dozens of flyspeck towns inbetween. Multicultural Australia is not just some big city phenomenon, nor is the bush just some Anglo-Saxon enclave, at least not down along the river.

On a Saturday evening on Dean Street in Albury, you can nurse a cappuccino, sip a glass of grappa, and watch the world float by. Young women in overtight jeans totter on white stilettos, arms outstretched for balance, cigarettes circling erratically, as admiring young men do laps in cars of chrome and polish, the doof-doof of their stereos syncopated with the low roar of their engines. It's a proud street, Dean Street, where the buildings have names—Temple Court, Corban's Building, Abikhair's Building—and dates—Mates Limited 1850, the Beehive Building 1888, Bogong Chambers 1923. Up and down the main street, people stroll in the late-summer warmth, emerging from Thai, Indian and Japanese restaurants, or perhaps from the old favourite, the Imperial Chopsticks Chinese. Perhaps they have been to the movies at the Regent or are on the way to the Commercial Club, where billboards advertise the imminent arrival of Gerry and the Pacemakers. Or perhaps they are simply promenading to the Paris end of Dean Street, where the Botanic Gardens, Albury's own Tuileries, stand behind a large sign advising that their verdure is due entirely to bore water.

And as you sit with your coffee and digestif, you can reflect upon what a remarkable achievement the river has been—an inland sea of stepped weir pools reaching from Mildura to the ocean, the river stretching above Mildura all the way back to Hume Dam, a river that never runs dry even as it traverses the arid interior of the driest habitable continent. And sitting above it, the Snowy Mountains Scheme, not just Australia's greatest civil engineering scheme, but unwittingly part of its greatest social engineering scheme. And as you sit on Dean Street, for a few minutes you might forget the environmental cost, hold in abeyance the impact of Aboriginal dispossession, and allow yourself to marvel at the sheer scale of the achievement. They did it: when they found the inland sea did not exist, that Matthew Flinders had got it wrong, they simply turned around and built it. In a dry and drought-racked land they built a 2200-kilometre-long guarantee of water, where pleasure boasts and paddle steamers could loiter year-round, where farmers could entice abundance from desert sands, and where the towns of a new Australia could flourish in the sun. Sitting in Dean Street, it makes the Opera House and the Harbour Bridge and the MCG and the new Parliament House and all those other capital city icons appear nothing more than baubles.

And so I raise a silent toast to all those unknown souls who built it. As I drain my grappa, I wonder if this is it, if I am seeing it at its apex, in the moment before it starts its long, slow decline. I think of Dartmouth Dam, of Jan Liepkalns who came from Latvia to help build it, of his son Peter who helps manage it, and of the blotting-paper catchment that refuses to let go of what little rain chances to fall within it. I think of the clear and pure stream that Cameron and I found in the mountains, the stream that tasted so cold and so sweet. And I wonder if it can endure, this great artery of Australia, this magnificent fabrication of a river, or if the mountain forests will burn and burn again and the great river will blow away in the gathering dust of climate change, and whether we will even appreciate what it is we have built until we lose it.

6

THE FOREST

The sparse woodland of Barmah State Forest

The river west of Albury is full: full of water and full of people. It's Victoria's Labour Day long weekend, the second weekend in March. Officially, autumn is seven days old. Yet the heat, the light and the spirit on the river are of summer. Melbourne has emptied, its denizens escaping east and west for one last taste of summer by the sea, but up in the hinterland, the people have come in their tens of thousands to be by the river.

I stop at Corowa, 60 kilometres downstream from Albury, a town awash with tourists and locals alike. It may be the worst drought in 100 years, the global financial crisis may be blowing tumbleweeds down Wall Street, but on this sunny Saturday morning in Corowa it's not possible to find a car park along the main street. Instead I leave the Hyundai on a back road opposite the red-brick courthouse, largely unchanged from those two days in 1893 when it hosted a pivotal constitutional convention. A brass plaque records the occasion for posterity. On a fence post nearby, some youth has scribbled a more contemporary message: 'fuck the police'.

At the caravan park down by the river, tents and caravans and motor homes have spread out like some exotic weed, covering every available inch of ground. Kids are riding bikes, playing cricket, pushing toy cars through the riverbank sand. One group is taking turns swinging Huck Finn–style on a rope out above the river; they hang for a moment at the apex of the swing, let go, and tumble with a yelp and a splash. Adults are basking in collapsible director's chairs beside limp fishing lines, or are strolling along the riverbanks or yelling half-heartedly at squabbling offspring. More than twenty years ago I spent an indolent afternoon here with my friend Andrea. We'd been driving aimlessly round Victoria, sodden with time but parched of purpose, and had found ourselves becalmed beside the river in the heat of a summer afternoon. There were no crowds that day; we'd had the river more or less to ourselves. We'd frittered away a couple of hours swimming out into the current, rolling otter-like onto our backs, and letting the river float us a kilometre or two downstream. Then we'd clamber out, return along the bank and jump back in. It's a dim memory now, as these things are, but I still remember the luxury of it, the sense that time was endless; all one needed to do was climb out, walk back and do it all again.

I can't say whether it's the memory of that day, or the mounting heat, or echoes of old desires, that lures me from the highway as I continue west from Corowa. The river red gums have been advancing and retreating in a line off to the south, whispering to me through the passenger-side window, telling me the river is close. I see a sign, 'Collendina State Forest', and on a whim I pull off onto a dirt track. I manoeuvre the car down towards the river. And there it is: wide and deep and alone; the campers have left me my own little piece of solitude. I strip off and edge down the high bank, the dry grass of summer sharp beneath the soles of my city-soft feet, the sun aggressive on my pale skin. The surface is an olive green, shimmering in the sun. I can see the mid-river current moving rapidly, rippling its muscles beneath the surface, not something to be taken lightly, not when swimming alone, even on a day as fine and cloud-free as this one. I ease myself into the water. It's cold enough to turn my breath shallow, even at the culmination of a long hot summer. It's the deep water from Lake Hume, released for the irrigators to finish off their crops. I dive under, revelling in the chill against my skin. The deeper I dive, the colder the water. Deep down, down near the bottom, I open my eyes. There's not much too see—a few centimetres of translucent green, the hint of a golden surface above—but the water is soft against my eyes, with none of the stinging chemicals of the swimming pool or the hard cleanliness of the sea. Resurfacing, I feel embraced by the river. Emboldened, I swim out from the still water, past the protective lee of a small island, and feel a thrill as the current takes me and races me off downstream. I want to dive down again, dive deep, and run sleek underwater with the current, feel it propel me forward. But I can't take the risk: a fallen tree, an unforseen snag, the pressure of the current coming in behind me; drowning could come far too easily. I'm older now, more cautious. There are children, responsibilities. So instead I stay on the surface, taking pleasure instead from the forest wheeling past on the far bank, a deep-green wall where cockatoos screech. I pass the end of the island and swim back to the bank. For the few strokes it takes, my shoulders revel in the movement. The river mud oozes under my feet and between my toes. It's been too long.

I've always had an affinity with rivers. As a baby, a toddler, a small child, weekends were spent beside rivers and, unintentionally, in them. My family moved to Canberra in 1961, part of the Menzian tide, back before Lake Burley Griffin was filled. Of a Sunday we would team up with another family, the Stacks, and head out to picnic beside the surrounding rivers—the Cotter, the Molonglo, the Murrumbidgee—where I would inevitably invent some new way of falling in. I never did it intentionally; it would just happen. I'd scramble after my elder brother, an adept climber of trees and leaper of rocks, and I would somehow slip and end up soaked through. It became a standing family joke, and my mother would pack a spare set of clothes in anticipation. My brother, enamoured with the emerging legend and keen to perpetuate it, even as I grew older and more sure-footed, would contrive to push me in. I would bawl my protests, but my mother would simply unpack the spare clothes and say, 'There, there, dear. There, there'. Mum, who had grown up by the beach in Melbourne, was deeply suspicious of rivers. And so, while we would barbecue beside them, apart from my own idiosyncratic immersions, we would seldom swim in them. My childhood was spent in swimming pools and, occasionally, the sea, but rarely in rivers. Nevertheless, my barbecue baptisms must have left some impression: the first real book I read, aged six, was *The Wind in the Willows*. I can still conjure in my mind's eye the book's cover, the EH Shepard illustration of Ratty and Mole in their boat, adrift on the river.

Much later, in the last two years of high school, a group of us would take a day off and head out to the Murrumbidgee, where we would lounge around naked—swimming in the river, lolling about in waterfalls, smoking cigarettes as we sunbaked on the sand. We had a secret spot, reached by a hike across sunburnt paddocks, the 'no trespassing' signs ignored. Murray, Ben, Rosie, Margaret-Ann, Carol and I, idling the afternoons away, while a current of sexual frisson flowed as we surreptitiously checked out each other's bodies. I should go back there one day, just to see what it's like, but I'm somehow reluctant. For our secret spot is secret no longer: Canberra has sprawled out to the edge of the river—'the Murrumbidgee corridor' they call it—and people go jogging there now, and walking their dogs and, hopefully, swimming. It was the ancient Greek Heraclitus who likened history to a river, observing that no man could ever step into the same river twice, for both the man and the waters will have changed.

Not far below Collendina State Forest, the crowds impose themselves again. At first it's just a fisherman or two, pulling off towards the river, towing their boats behind them. And then I see why, as the river spreads towards the road. A lake appears, lapping at the feet of drowned trees. Boats float among the skeletons, lines cast. It's Lake Mulwala, backing up behind Yarrawonga weir, and it's big.

Yarrawonga is different to the weirs on the lower Murray. The weirs were built from Mildura and lock eleven down to the barrages (effectively lock zero) at the insistence of South Australia, to guarantee navigation. That's not to say irrigation water isn't pumped out from the weir pools; it is, a lot of it. Of the remaining weirs/locks of the twenty-six originally planned, only the one at Euston (lock fifteen, 232 kilometres upstream from lock eleven) and the one close to Echuca (lock twenty-six, a further 538 kilometres towards Albury) were built. None of the nine navigation weirs intended for the lower Murrumbidgee were constructed. Instead, irrigation weirs were built: five on the Murrumbidgee and the Yarrawonga weir on the Murray. The weir at Echuca also functions as an irrigation weir. The purpose of these weirs is to raise the levels of their ponds high enough so that the water gravity-feeds through a series of canals out across the wide plain. The ability to do this and the flatness of the land was the magic combination that permitted huge tracks of land to be opened up to irrigation—not just for horticulture, but for broadacre irrigation like rice, cotton and dairy. Pumping water is expensive: not just running the pumps, but the costs of buying and maintaining the pipes and other infrastructure. So where water is pumped, irrigation tends to be more intensive, with permanent plantings spread over a smaller area. But where the water can flow under its own weight through open-air canals and ditches, the costs of conveyance are much lower and broadacre irrigation becomes more attractive. And so in an average year, back when there was such a thing, Yarrawonga diverted about 17 per cent of the Murray's flow, sending it into Victoria via the Yarrawonga Main Canal, and into New South Wales via the larger Mulwala canal, spreading it over more than 8280 square kilometres, or an area roughly the size of greater Melbourne. Up on the Murrumbidgee, the Berembed and Gogeldrie weirs gravity-feed the Murrumbidgee Irrigation Area's 3624 square kilometres, an area twice the size of Adelaide. The water, captured and stored in the Hume and Dartmouth dams, comes down the river, is diverted by the weirs, and flows out through the canal and ditches onto the farmer's fields. What could be easier, or more efficient?

I drive through Mulwala and out across the lake towards Victoria, the long low bridge twisting and dipping and only just meeting in the middle, the result of an interstate planning imbroglio. Yarrawonga appears, and I don't know what to make of it. I've journeyed in anticipation of droughts and climate change, of struggling irrigators and stressed environmentalists and uncertain compromises, of river communities and their histories, of a slow brown river oozing out across the great western plain. Instead I find myself in a boomtown, a Gold Coast on the Murray. There are no highrises, to be sure, but there is the unmistakable atmosphere of a holiday town: motels and caravan parks and cream-brick holiday units; cafes selling fish and chips and pizza and ice-cream. Villa Crystal, Casa Bella del Lago and the Cyprus Lagoon Apartments are all booked out; the Mulwala Paradise Palms Motel, Capri Waters Country Club and the Cool Waters Caratel have no vacancies. Not on this holiday weekend. Seagulls wheel and squawk by the lake, scavenging chips from families picnicking on the grass beside a skateboard park. An enterprising young man has parked a barge at the water's edge, music pumping out of huge speakers, enticing the skateboarders to buy his wares: cold soft drinks and packets of potato chips. On the lake itself, powerboats roar, towing waterskiers. A couple of lairs on jet skis perform the aquatic equivalent of burnouts and doughnuts. A little further along, families have gathered at a swimming pool, not one of the antiseptic pools of suburbia but one built into the waters of the lake itself. It reminds me of the sea baths that dot Sydney Harbour, complete with boardwalks and grills corralling off a patch of water. It's somehow unexpected out here: the Dawn Fraser pool at Balmain without the breaststroking pensioners; the Redleaf pool at Double Bay without the lobster-red young men from Britain; but a similar feeling nevertheless. A water-slide elicits squeals nearby. Infected by the seaside atmosphere, I buy fish and chips. Three hundred kilometres from the ocean, they don't taste half-bad. I've had worse at Bondi.

I finish my fish and chips and go for a walk around the lake, across the weir. I've never quite worked out the difference between a weir and a dam. The mighty concrete and earth constructions of the Snowy Mountains Scheme are clearly dams, just as the low structures built within the channel of the Darling are clearly weirs. It's the ones in-between I find confusing. One definition depends on function rather than scale: a dam holds a reservoir while a weir is simply there to divert or regulate the flow. But then there's Lake Hume, clearly a reservoir, backing up behind a structure that is sometimes referred to as Hume Weir and sometimes as Hume Dam. At Yarrawonga the weir is an imposing concrete edifice, larger than Scrivener Dam in Canberra, which holds back Lake Burley Griffin. It's called a weir because it diverts water for irrigation and regulates the river's flow. But Lake Mulwala is a substantial reservoir, not some piddling weir pond. So I'm still confused. I look down from the height of the dam wall, watching as a boatload of insubordinate fishermen push their craft beyond the safety buoys and cast their lines into the churning flow below the fish ladder.

I lived by the Fish River in 1984, my last year at uni in Bathurst. My friends Lynne and Diana and I shared an ancient homestead, Westham, built in 1830, the same year Sturt headed off down the Murrumbidgee in search of the inland sea. I've heard that a prominent Sydney businessman has restored the old house at great expense, but back when we lived there, its glory had collapsed into calming dilapidation: the plaster ceiling sagged, wild roses grew through gaps in the sash windows by the fireplace, possums scurried in the roof. On the veranda, out through the French doors, a century and a half of paint had eroded into a patina that, in the rays of the setting sun, was as subtle and evocative as any deliberate work of art. The old farmhouse cost us $20 per week to rent, an outrageous mark-up on the $10 our predecessors had paid. I spent a lot of time down by the Fish River that year, ploughing through the books for my literature subjects—
The Fortunes of Richard Mahoney, *Return to Coolami*, *Capricornia*,

Coonardoo—while all around the landscape rebounded from one of eastern Australia's most devastating droughts. I read the books by the banks of the river, or beside a small creek up behind the outbuildings, or in front of Westham's prodigious fireplaces. In summer we would venture further upstream, up beyond the O'Connell pub to swim at Flat Rock. One day, my mate Tim and I borrowed a canoe and tried to paddle from Flat Rock to the pub. It was an early lesson that blue lines on a map are no guarantee of water. We ended up carrying the canoe—over rocks and through walls of tea trees and round black-berry infestations—more than we paddled in it. Nevertheless, we were given a heroes' welcome when, Sturt-like, we eventually emerged from the wilderness and pulled in beside the pub's beer garden.

It's a great metaphor, the river. Throughout history and across cultures, it's been a symbol of life itself (although an old friend vehemently argues that cricket presents the more accurate parallel). The Hindus on the Ganges, the ancient Egyptians on the Nile, Babylon in Mesopotamia: civilisation grew from agriculture and, more often than not, agriculture grew from the rivers. It was humankind's constant companion, in flood and drought. In the Bible, in Revelations, Saint John recounts a vision of heaven: 'And he shewed me a pure river of water of life, clear as crystal, proceeding out of the throne of God and of the Lamb ...' Jesus, himself baptised in the River Jordan, is reported in the Bible as saying: 'All the rivers run into the sea; yet the sea is not full; unto the place from whence the rivers come, thither they return again'. It's not just rivers, of course. All aspects of the landscape have accrued their own symbolism: the mountains represent spiritual purity; the desert, despair and desolation; the forests, narrow paths through unseen perils; and the ocean, the profound and unfathomable. But it's the river that links them all: rising in the mountains, growing strong and purposeful in the foothills of youth, navigating narrow forest gorges, slowing across the meandering plains of middle age, before eventually mingling with the unknowable ocean. It's not just me; we're all of us soaking in a metaphorical river.

But it's an old metaphor, based on a natural river, the river of the *Oxford Dictionary*. Not all rivers run to the sea, not anymore. Not in Australia they don't, not during the worst drought in a century. And so I wonder, what sort of metaphor do we derive from our modern rivers: rivers that rise in the concrete diversion channel of a dam; that are measured and regulated their entire length; that are split apart naturally and unnaturally, diverted to soak into desert fields; that have their salt pumped out to evaporate in shallow pans; that are bought and sold; that are corrupted and polluted and degraded; and that, ultimately, fail to reach the ocean? A river that doesn't reach the sea— what sort of metaphor have we created for ourselves? Have we lost our most potent parallel or, more disquieting, have we instead fashioned one that perfectly reflects our modern secular lives?

I set out to travel along Australia's rivers and instead I find myself travelling through a metaphor, my own and my country's. I've been to the mountains with Cam and I've been to the desert with the ghost of Henry Lawson. It's time to visit the forest.

Drive north from Echuca towards Deniliquin and something subtle, almost imperceptible happens to that hot dry plain. Ever so gradually, it rises. Although I've come looking for it, it's nevertheless difficult to be certain if the ground truly is tilting upwards or whether it's just my imagination. To the west, where the setting sun is casting the fields in gold-fringed relief, no sign can be detected. The first confirmation comes from the east—stunted trees begin to appear in the distance, yet as they move closer to the road, I realise they're no shorter than normal; something peculiar is happening. I pull off the road near the old Moira Station homestead. The trees are close now, just 100 metres away, yet all I can see is their crowns.

I'm standing atop the Cadell Tilt, a rift in the earth's crust. The lift in the land, so slight as to challenge the perceptions if approached from the north, south or west, has in fact created an abrupt north-south escarpment 50 kilometres long, a fissure in the earth, readily apparent if approached from the east. It's not high, varying between 8 and 15 metres, but high enough to dam Australia's greatest river, to spawn the world's largest red gum forest. The road runs along the raised plain, dry and summer brown, while off to the east, a great woodland stretches to the horizon.

Sometime between 13 000 and 25 000 years ago, the Cadell Fault raised this long slab of land along a sharply defined north-south line, leaving the land to the west up to 15 metres higher than the land to the east. The present-day Cobb Highway runs atop it between Echuca and Deniliquin. The tilt created a massive natural dam, stopping the Murray River and forming a huge lake. For hundreds, probably thousands of years, the dam held. The river eventually found its way around the obstacle, but only with difficulty. To the north, the water was pushed up the Edward River, flowing past modern-day Deniliquin, before rejoining the Murray a couple of hundred kilometres west, downstream of Swan Hill. Along the way, the flatness of the plain splits the Edward into the Wakool River and the Wakool into myriad creeks. For the most part, they come together again and eventually reunite with the Murray. To the south of the Cadell Fault, the Murray started pouring into the bed of the Goulburn River, down by modern-day Echuca. In geological time, the stretch of riverbed between the Cadell Tilt and Echuca is brand new, by some measures only 500 years old. It hasn't had time to carve a deep channel in the hard rock of the plain, so instead it flows in a shallow stream across its surface. This is the Barmah Choke. The choke can only carry a limited amount of water, about 8500 megalitres a day. The ancient dam has been breeched, but not entirely. This frustrates the hell out of the authorities who want to send more water down the river. For decades engineers have been advocating the construction of a canal to circumvent the choke. But the obstruction has created one of this country's natural wonders. For if too much water comes down the river, it floods out over the riverbanks of the choke, feeding a massive wetland. Over thousands of years, the annual flooding of this wetland created the world's largest river red gum forest: Barmah-Millewa. I stand atop the Cadell Tilt and look out across the tops of the trees, and once again admire the perversity of Australia's rivers. Whereas the foreign archetypes have their deltas at their mouths, the Darling has one at its beginning, and the Murray inserts its own delta halfway along itself.

Sharon and Russell Terry live beside an irrigation canal 15 kilometres south of the Murray. Their home is an old water bailiff's cottage. The bailiffs are being phased out now, but for generations they were the guardians of Victorian irrigation. Their job was to supply water to irrigators, opening and closing the gates between canals and farms. The canals snake out across the land, following contours rather than logic, and the bailiffs have travelled them, dispensing the water and keeping the accounts. It's been a haphazard affair, with planks of wood inserted as leaky regulators and Dethridge wheels measuring flows; the wheels are notoriously inaccurate, but only ever in the farmer's favour. The bailiffs also monitored the system for leakage. And theft. In recent years they've been rebadged as Water Distribution Officers, but the irrigators still call them bailiffs. Good for them. What a word: 'bailiff'. So evocative, so accurate. It's like the Barmah Choke. I wonder what Orwellian wonder-words the bureaucrats may soon extrude to replace it: the Mid-Murray Flow Attenuator? The bailiffs are being replaced by computerised flumes and electronic meters, a much-needed reform, but the end of an era nevertheless.

Russell Terry stills works for Goulburn-Murray Water, but he's no bailiff. Instead, he's been seconded to Future Flows (marketese, the insidious private-enterprise cousin of bureaucratese), a consortium modernising the local irrigation system. But 51-year-old Russell, his greying hair cropped short around a matter-of-fact face, is not your average water worker. He's planted the 1.6-hectare block surrounding the couple's cottage with native trees and bushes, turning it into an open-air aviary. He tells me he grew up in Melbourne, part of a tough-as-nails clan, with cousins on the docks and in the meatworks and in and out of prison. Yet from an early age, Russell felt an affinity with the world of nature.

'I grew a native garden in my father's front yard. I was about fourteen or fifteen, and we were planting it out and putting in these young mallees, these eucalypts, in the front yard. And old people would come past, who had things like ash trees and willows in their yards, and abuse me, saying I was a dumb kid putting those eucalypts and gum trees in. They said the roots were going to rip up the footpaths, that they were going to create havoc. It was anathema. It all had to be European or it was going to do some damage to the suburbs and the houses and the sewerage lines and such. And all the time these people were putting in willows. I had a cousin who was a plumber. He made a fortune clearing willow roots out of sewer pipes', Russell chuckles, as the three of us drink tea in the couple's kitchen.

'I always wanted to move to the bush. Somehow it was in my blood. I was a bit of a yobbo as a kid. I kept birds, skin dived, went hunting and fishing. Later, I ended up going deer stalking because I couldn't bear to kill anything native. Then I couldn't bear to do that, either. So I'd spend my time with a gun over my shoulder, pretending I was deer stalking, when in fact all I was doing was taking photos of the bush and going up these beautiful gullies and hiking around.'

Russell's wife, Sharon, is another black sheep, but from a different type of family, a country family. She grew up at Robinvale, down along the river near Mildura. She was the first of her family to go to university, earning a psychology degree. Now she's a youth worker at Mission Australia at Shepparton, 60 kilometres away. 'I was born on the edge of the Murray; we always spent weekends on the river. It's a really strong part of me: I have to live by the river. I feel lost if I don't live near the Murray', she says. Sharon has a friendly, down-to-earth manner about her, as if she's comfortable in her own skin. She has red hair, an infectious smile and protruding ears.

Russell and Sharon take me into the forest in their Subaru. The trees start abruptly, about 10 kilometres north of the Terry's cottage, with a clear line where farmland ends and woodland begins. At first glance the forest is not that impressive, just a stretch of dry, relatively open bushland. I'd been reading about this epic wetland system, about people paddling canoes from lagoon to water-lily-bejewelled lagoon. Instead, the country doesn't seem that different to the bushland near my house in Canberra, just flatter. The gum tress look like gum trees, the sparse dry grass like sparse dry grass. Where, I wonder, is the magic? As if eavesdropping on my thoughts, Sharon asks Russell to stop the car. We climb out beside a single-lane bridge. Sharon says this is Smiths Creek. It's dry, utterly dry. It reminds me of the Culgoa, and not just because it's devoid of water. It has all the hallmarks of a substantial waterway, with a deep concave channel and clearly defined banks. Russell says the creek is an anabranch, that it runs out of the Murray, not into it, as the choke raises the main channel of the river up to the height of the surrounding forest. He says Smiths Creek winds its way through the forest and back into the river below the choke, a small part of the Murray's mid-river delta.

'This creek, before they put a regulator on it, it ran crystal clear and it was more or less permanent', says Russell. 'Because of the choke, it hardly ever got below a certain level. The stories that creeks like this used to dry up regularly are crap. It happened about twice in recorded history. This flowed at a high level nearly all the time because of the slight gradient taking it off from the river above the choke. They say it was full of platypus, crays, the whole thing. Now there's a regulator to stop it, to keep the water in the Murray and out of the forest. It was one of the ways the flood spread through the bush. It would break out into all these streams and from there into the bush.' The regulator is a gate that has been placed across the creek where it flows from the Murray. It was put there about the time of World War II.

I ask Sharon about the state of the trees. She agrees they look pretty shabby, but says they were much worse a few months earlier, back before some decent rains fell in November. Before that, the foliage was so thin it was difficult to find shade. She points out a couple of particularly miserable specimens, with foliage restricted to just one or two branches. Other branches lie on the ground, discarded by some last-ditch survival mechanism. 'They'll be dead in another six months or so', she says. We're only 100 metres or so into the forest and already I'm revising my preconceptions, abandoning fantasies of canoe expeditions and Kakadu-like splendour, replacing them with a more desperate reality. Yet Barmah-Millewa is one of six environmental icon sites identified by the old Murray-Darling Basin Commission. It's an internationally recognised, Ramsar-listed wetland. Ramsar is not an acronym, as many people assume. It's the name of a town in Iran where an international treaty to protect wetlands was adopted back in 1971. Signatories to the treaty, including Australia, pledge to conserve the ecological character of wetlands listed on the Ramsar register. Barmah-Millewa's preservation is also supported by international migratory bird agreements with China and Japan. And it's sick. It hasn't had anything close to a proper flood since 2000. Before the Hume Dam, the choke ensured the Murray would flood the forest almost every year, for months at a time. It's not dying, not yet, but it's changing, from a wetland to a woodland—no-one can say exactly how, or what this means.

We drive a little further and my spirits are lifted by an explosion of greenery in a low-lying area beside the road. But Sharon disillusions me. The trees are indeed river red gums, but their growth in that location is unnatural. It's too low-lying. They've germinated because of rain, not flood, and are crowding out vegetation that previously grew there. Normally, regular floodwaters would have drowned the young trees; by rights they should be growing on slightly higher ground. The forest is askew. Even the floods, when they do come, come at the wrong time. Before the Hume Dam, they came with the winter rains and the spring snowmelt. But for the last sixty years they've come during summer, if at all, spilling over from irrigation flows; for the last decade, there has been precious little flooding, in winter or summer.

Nearby a forest giant stands dead, not killed by drought but by the forestry department. It wasn't cut down to be milled, just ringbarked and left to die. 'The theory was to get rid of the mongrel old ones to allow room for the young saplings to come up', explains Russell. 'Look at it. So old and gnarled, you can imagine just how many bats and possums and reptiles and insects it must have supported. And the forestry department was just going around killing them.' Robbed of my vision splendid, I'm starting to feel a bit glum. But not Sharon and Russell; they're full of hope. The state forest is about to become a national park. Grazing has stopped, the last cattle removed eighteen months ago. Logging should cease within another six months. They believe the forest has a second chance. We stop by a small fenced-off reserve stretching along a ridge dotted with yellow-box trees. The reserve was established as part of a research program some twenty years ago. Inside the fence, beside gnarled old bull oaks, young trees are growing. The forest is regenerating. And now, outside the fence, there are signs of regrowth despite the drought. Sharon points out small gold-dust wattles. She says the regrowth has come since the cattle left. She smiles, some of the enthusiasm spreads, and I don't feel quite so glum.

My outlook further improves when we reach the river, and for a fleeting moment, as we walk towards the bank, I catch the vestiges of the forest's elusive magic. It's been corralled into a narrow strip running along the river, a remnant of what the forest must have been like in summers past, after the spring floods had subsided. For along the banks, the forest is close enough to the river to retain its vitality. The river red gums tower, lofty and robust, 40 metres high or more, dwarfing the straggling pretenders further from the river. The giants grow alone, claiming their own space. The bottom of their trunks are covered by straggly grey bark, as if to protect them from floodwaters. Higher up, the bark peels away, revealing whitish wood streaked with grey. The trunks don't climb so high before dividing into well-spaced branches: there is no umbrella like canopy; the foliage is well dispersed as the tree reaches higher, and for all their size, the gums offer only intermittent shade. There are one or two dead branches, even on the well-watered trees close by the river. The gums are hundreds of years old, home to possums and sugar gliders and koalas, to bats and goannas and parrots. Some river reds have been dated at 1000 years or older. They speak of what was once widespread and suggest what may yet be saved.

The Murray itself is running high, up towards the tops of the banks, as if to mock the thirst of the surrounding woodland. It's always like this in summer, day and night, as the authorities push the maximum amount of water through the choke without spilling it into the forest. It's an artificial flow but a flow nevertheless, and some of the water is reaching the closest trees. But this is no wilderness, especially not this long weekend. Along the banks, there are people—solitary fisherman and their pup tents, groups of men with boats and four-wheel drives and well-tended piles of empty beer bottles, and multi-family tribes with tents the size of bungalows, adorned with showers and generators and gas barbecues. We stop by one such encampment, where a large Australian flag holds pride of place amid the motorbikes, the pushbikes, the kayaks, the boogie boards and the Eskys. Half a dozen kids are burling around on bikes and on foot, playing their own idiosyncratic game of chasings.

'Some of the locals camp here for most of summer', explains Russell. 'They leave their gear here. Motorbikes, motorboats, the lot. It's first in, best dressed. The locals come in, put their vans in, and go home again. You should see some of them: hot water and river pumps and TVs and generators and lights. A lot of them are dead against the national park.' I imagine they are. I wonder if that's what the Australian flag is all about, some statement of settler rights. Or maybe it's just a leftover from Australia Day, six weeks ago.

We stop for lunch further down the river at 'the Gulf', where the river winds back on itself in a tight bend. Sharon has packed ham sandwiches and they taste somehow better in the shade of an old-man gum tree. In the forest, by the river, the persistent heat of the Indian summer has been subdued. It's a beautiful spot. Sharon and Russell enthusiastically point out the different varieties of bulrushes and reeds growing in the river shallows, telling me that these too have rebounded since the cattle were banished. Another two streams run off the river at the gulf, but like Smiths Creek, regulators have blocked their flows. Russell tells me that long ago there was a punt here, back in the nineteenth century. The Gulf is also where a fugitive named Ned Kelly crossed into New South Wales, heading for Jerilderie and history.

While we eat, I ask Russell about the forest, about what first attracted him to it. 'It must have been thirty years ago', he says. 'I was blown away. It was in much better condition, of course. It smells different than other forests, a unique, beautiful smell. It has a feel that's almost exhilarating. It looks strong, it has a strength about it. The third time I came here, it was in flood. We went canoeing through the trees. It was the stillest water I have ever seen. It lay there, in the forest. It glistened like a mirror, and yet was somehow clear as well. We were in a canoe; we followed some swans. We could see their feet paddling through this clear, still water …'

Russell's passion is birds. He knows them by sight, knows their habits, knows their needs. And for many, the greatest need is water.

Not the relentless flow of the Murray or the sterility of irrigation channels, but the still, flat lagoons of a flooded swamp. 'It's devastated the wetland birds. There are some birds that can live in the irrigation areas, like ibis or wood ducks, but the ones that only breed in the swamps—blue-bill ducks and the plumed egrets, the darters—they need a flooding to a have a proper nesting event. My fear is the numbers are crashing. Things like the glossy ibis and brolgas, they're pretty well extinct around here now. There were southern brolgas nesting around here. I've seen the odd one about the place, but there is no breeding at all … Grey teal was a fairly common bird, but they nest in swamps, not channel systems. I haven't seen a breeding event for them for about ten years, I reckon. The last time it flooded properly the teal nested in here by the tens of thousands, but you barely see a single grey teal anymore. The same with the royal swan, their numbers have crashed. There were over 2000 nests here back in 2000.'

Back in the car, we drive in silence. Russell and Terry oscillate between mourning the damage done by the lack of water, and enthusiasm at the promise of the new national park. We stop at Thistle Point, close by the choke itself. The river here is full, right up to the top of the bank. Indeed, the banks are higher than the surrounding forest, small levees built up naturally over the years. The river is flowing across the top of the ground, held by the fragile levees, not through a channel. This is the youngest part of the Murray, perhaps only 500 years old, a blink of geological time. It wouldn't take much more flow for it to spill out and into the forest. These minor floods happen from time to time, courtesy of a big rain, or when too much water comes down the Ovens and the Keiwa to be contained by the Yarrawonga weir, or when the authorities sneak the forest a quick drink. But a proper flood, a forest-wide flood, would require a deliberate decision by government. It would need water sent from Hume Weir and passed along by Yarrawonga, the regulators opened on Smiths Creek and all the other feeder creeks. But that's not going to happen, national park or no national park, for as long as the worst drought in a century continues and communities the length of the river cry out for water.

⌣⟶

A few miles south-west of the forest, out among the farmlands, there's a low ridge, a continuous hill flowing like an unbroken wave across the plain. It's maybe 30 or 40 metres high and covered in golden grass. Russell and Sharon show it to me. At first it doesn't appear remarkable, but the more I consider it, the more unusual it seems. It's all alone: a single unbroken undulation frozen on the otherwise flat plain. Nor does it run dead straight but rather tracks in a smooth curve, a huge crescent. It is both strange and familiar. The couple tell me it's the shoreline of an ancient lake, elevated above the surrounding plain by centuries upon centuries of wave and wind action, similar to the elevated shorelines ringing Lake Menindee and Lake Parmamaroo. It seems that the Cadell Tilt, lying a few kilometres further west, did more than simply dam the Murray; it dammed the Goulburn as well, forming a huge lake, an inland sea: Lake Kanyapella. For thousands of years it covered something like 400 square kilometres, stretching from modern-day Barmah to Echuca. This long, golden wave of earth is its last remnant. It makes me think of Charles Sturt and his failed quest to find the inland sea; he wasn't wrong, he was merely late.

The ancient shoreline is a reminder of just how dramatically the landscape can change, how impermanent our waterways can be. According to the most extreme of the climate change predictions, this region will itself be dramatically transformed. If they're right, Barmah-Millewa won't survive the century, not without intensive intervention. South-eastern Australia is forecast to grow drier and drier and the inflows into the Murray to grow weaker and weaker, until the river will not carry enough water to spill out across the plain and into the forest, choke or no choke. It's the great unknown: is this merely another drought, or is it the harbinger of climate change? Throughout the basin, farmer after farmer has reiterated to me their faith: this is a drought, a terrible drought, but a drought nonetheless. A familiar part of a familiar cycle. They pull the tattered evidence from desk drawers and bookshelves. Old ledger books, pages yellowing, spines crumbling, with handwritten records of rainfall. They flip the pages: back years, back decades, back generations. The handwriting changes. Felt tip preceded by ballpoint, ballpoint by pencil, pencil by copperplate. Their writing, their father's, their grandfather's, their great-grandfather's.

They brandish the evidence, the catechism, that demonstrates this drought is no worse than the great droughts of the 1960s, the 1940s, the 1890s. They are right—who can say otherwise?—when they say the lack of rain is no worse than in decades gone by. But they are also wrong. The rain falling on their lands may supplement the Murray River if it can circumvent the contour ploughing and on-farm dams, but it won't determine the river's flow. The water in the Murray comes from the mountains, the water that once flowed plentifully into Dartmouth and Hume. And the ledgers at Dartmouth—computerised, analysed, scrutinised—tell a different story. This drought is different. It's worse.

The science of climate change is complex, rapidly evolving and difficult to comprehend. It's not easily held to account: the projections keep changing, the variables shifting, the absolutes few and far between. Inevitably some of it will be wrong and much of it will be modified. But break it down into its component parts and it becomes easier to understand. The first effect of a warmer atmosphere is simple: if more energy is fed into a dynamic system, that system will itself become more energetic, more dynamic. It's a Newtonian fact. Hotter hots, drier dries, wetter wets. Maybe even some colder colds. The real-world consequences are larger and more frequent storms, cyclones tearing in from the Coral Sea to ravage Queensland's tropical coast, with more and more rain in the north.

The second effect of climate change, and this is the one that bodes ill for the Murray River, is that a warmer atmosphere is predicted to push Australia's weather systems further south. The Asian monsoon will penetrate further into the continent, bringing more rain to Queensland and the Northern Territory. In turn, the drier systems below the monsoon will be pushed further south, bringing less rain to the south-eastern and south-western corners of the continent. In the Murray-Darling, there will be less rain in Victoria and South Australia; less rain in the catchments of the Murray, the Murrumbidgee and the rivers of Victoria. Already, climate scientists are talking about something called the subtropical trough, another invisible line, this time drawn high in the atmosphere and stretching between Canberra and Adelaide, a line that hovers above Conroy's Gap. North of the line, the weather will either be wetter or, at least, no drier. But south of the line, drought will become more frequent, more prolonged and more punitive. Such a scenario could see western Queensland awash, Coopers Creek and Diamantina in flood, and Lake Eyre filling, while to the south and east, the Murray and its southern tributaries gasp for water like beached fish. The central desert could bloom while the Alps burn; there'd be bigger, more frequent fires. No-one will say for sure that climate change has arrived: such a call is impossible. No-one can say what is drought and what is some more profound and permanent variation. But what can be said is that what we have witnessed across eastern Australia this summer, and the summer before, and the summer before that, looks very much like how climate change is expected to look.

The science predicts other, larger effects, more dramatic in their impact, less certain in their eventuality. The third effect of climate change will occur if the great icecaps of Greenland and Antarctica melt. The icecap in the Arctic has been steadily receding, and so have parts of the Antarctic sea ice. Scientists say the process appears to be accelerating, especially in the Northern Hemisphere. Yet at this stage, most of the melt involves sea ice, which means no change in sea levels. The crisis will come if significant amounts of land-based ice also melts and runs into the sea. There is so much ice that sea levels could rise by metres. Combined with larger and more frequent storms, this could spell disaster for Australia's coastal cities. And then to where will the refugees from Sydney and Melbourne retreat: to the emerging desert of the Murray-Darling Basin? Where will the refugees from the desert of the basin flee: to the drowning cities of the coast?

The fourth effect of climate change is the great unknown. At some stage, some mechanism may reach a tipping point and subvert major weather systems. The Gulf Stream might switch off, El Niño might become permanent, the monsoon might radically alter its nature. No-one can say. These phenomena are essentially unknowable because no-one really understands how all the weather systems of the world interact. Even if there were a computer powerful enough to process all the information, no-one would know what information to feed into it, how to collect the data or how to weight it. But for now, it's the second effect that's exercising the minds of those who debate these issues within the corridors of power and in the halls of academia. Are the weather systems really moving south? Is drought destined to become a permanent characteristic of the Murray and the other rivers of the south? Will the river red gum forests of the southern flood plains endure?

⌒

Tim Mannion is seventy-six years old. He's a hearty and healthy looking bloke, a big strong man after a lifetime of farming, with a full head of hair and a smile boasting mostly original teeth. Dressed in a flannel shirt and blue overalls, he greets me with a knuckle-crusher handshake. But his vitality belies recent travails, including a cancer scare and a broken hip. Tim was born a stone's throw from the Barmah forest and has spent his life along its edges and under its canopy. Even now, from his home in Nathalia 25 kilometres south of the forest, his bushman's senses remain keen. He reckons if a fire breaks out in the forest, he can determine its approximate location from his front door. A phone call to the old bushman Gerry Moor at Barmah and the two men can triangulate the fire's position on the maps they carry around in their minds. Another phone call will tell the fire fighters precisely where to send their crews.

Sitting at his dining table, Tim tells me the natural forest, the red gum wetlands created by the Barmah Choke, is long gone, and not because of drought. He says what remains is the result of more than a century of European management, shaped by cattle grazing, logging and the managed flows of the river. He believes the national park is a mistake, that the forest requires ongoing management if it is to survive and remain healthy. He says the forest was exploited from the moment settlers arrived in the early nineteenth century, and this was formalised when the old forestry commission took control around the beginning of last century. The local Aboriginal people left the forest, taking their traditional burning regimes with them. Tim tells me that as the pressure to settle land close by the forest grew, a levee was built around its southern edges, protecting prospective farmlands from flood, and having the unintended effect of containing the waters within the forest, raising them and making them longer-lasting.

'The levee was built with wheelbarrows. The workers were paid so much a yard. Whole families camped along the riverbank. And they'd pay local farmers to plough the ground for them and then they shovelled it into wheelbarrows and ran it up a plank and onto the levee bank. The bank was about 5 feet high and 34 miles long. They did it in fits and starts. They had a break for a while; there was a gold rush in Western Australia and fellows left the job to go chasing gold, and then they were unemployed again and came back to it. They got a shilling a yard for building the bank with their wheelbarrows. And fellows made money. If they had a family, then the family could all help. They could all shovel dirt into the wheelbarrow.'

The levee defined the forest, allowing settlers to select farmland right up to the bank, land that had previously been swampy and flood prone. Systematic logging quickly followed. Sawmills were established along Smiths Creek, where the permanent water kept the steam engines running. The first wood came from the trees cleared from the new farm blocks. The mills turned the river red gums into lumber and sold it back to the farmers to build their houses, sheds and fence lines. But once the local market was satisfied, the mills moved away from Smiths Creek towards the river. The annual winter and spring floods, contained within the levee, would make the forest impassable for months at a time, and the river was the only route to the outside world. The loggers would use the dry months of summer and autumn to fell trees, hauling them to the mills. The mills would cut the timber year-round, stockpiling it by the river. When the floods came, the paddle steamers could bring barges up the river and the timber would be floated down to the railhead at Echuca. The logging camps, cut off during the floods, were their own little communities. Tim points out the location of the some of the old mill sites on a map: Black Engine Lagoon, Green Engine Camp, Red Tank Lagoon. The map is redolent with names: The Big Bog, Waiting Plain, Buck's Sandhill, Doctor's Swamp, Little Rushy Swamp. There's the Steps, where a family cut steps down to the river to help their crippled father gain access to the water. There's Dinny's Dip, where a sleeper cutter, weighed down by his tools, fell into deep water and almost drowned. And there's Chinaman's Creek, where a cook from one of the logging camps wasn't so lucky. According to Tim, most of the camps had a Chinese cook, versatile and ingenious men who grew kitchen gardens, caught fish and shot kangaroos to keep the men fed through the months of isolation.

Another industry followed the loggers: sleeper cutting. The river red gum wood, hard as rock and water resistant, provided the foundations for the railways snaking across the landscape. And not just within Australia. The railways of British India were being laid on the hard dark wood coming from Barmah. The streets of Melbourne were paved with 6-inch blocks, stood vertical and bitumened over. They're still there, virtually everlasting, cushioning the traffic along the city streets, unseen under the car wheels and the tram tracks.

World War II came, first emptying the forest of men and then refilling it. Australia, with no domestic oil industry, was dependent on petroleum imports that were being sunk with sickening regularity by U-boats and German raiders. Petrol rationing was imposed and an ingenious alternative was developed: charcoal burners. Charcoal was placed into containers and heated, giving off enough flammable gas to power a car. It wasn't the most convenient source of energy, nor the most powerful, but sixty years ago Australians were using a fuel that was plentiful, carbon-neutral and renewable. Motorists would pull into service stations and instead of filling up with petrol, they'd buy bags of charcoal. The new industry brought new men to the forest, different men.

'There was a manpower shortage because all the men of that age were in the army or in vital jobs', Tim Mannion recalls. 'So it became the job of internees, Italians or other foreign nationals. They put them in camps. They had to work under supervision or report to the police once a week. There were old fellows who were too old to be in the army, and anybody who thought they could earn a shilling. There were even prisoners of war. A fellow who was burning charcoal in the forest would be allocated a couple of prisoners to work for him. He had to look after them and they worked for them. Once it got going a bit, people were coming from Melbourne, all unknown in origin. Once you'd been working on the charcoal for a week, you were black. It got into the pigment of your skin, and you never got clean again until you left the job. And so a fellow who was working on the charcoal was fairly hard to recognise. And before very long there were deserters and fellows who were trying to avoid the draft. All those fellows were working on the charcoal. They were camped in the forest, those fellows, and some even had their wives and families with them. They were all my friends. I was fifteen when it ended.'

At the age of eleven, Tim Mannion was thrust into this strange forest of black-skinned deserters, draft dodgers, prisoners of war, internees and fugitives. The great drought of the 1940s had taken hold, turning the family farm to dust, and Tim was sent into the forest for six weeks over summer to protect the family cattle. The cows were in danger of getting bogged in the drying forest waterholes, so he'd keep them walking about in search of feed, watering them in the river itself where they wouldn't bog. He'd settle the cattle where there was enough feed, rush home for something to eat, and then return to the forest to sleep with the cows. And as he walked the forest, he got to know the men, and they got to know him.

'Leo Bresnan, he was a terrible cook. My mother used to feel sorry for him. We used to kill our own meat but we didn't have much refrigeration, so my mum would say, "Take a bit of this down to old Leo Bresnan". "Ah, fresh meat", he'd say. He'd knock off work to cook it. He'd cut it into narrow strips like bacon and fry it up with potatoes. And he'd have it just right, and he'd say, "You can't eat good tucker without condiments", and he'd get a pinch of ash out of the fire and sprinkle it over it. He was a bloke who liked his charcoal', says Tim.

'There was a bloke charcoal burning on Black Engine Lagoon who had a son the same age as me, and we used to put in a bit of time together when I had the cattle settled down at the Black Engine sandbar. Lenny and I would go foraging through the bush chasing native cats and things like that. He had a flying fox to get water out of the river. He had two 10-gallon buckets on it. You'd get into the empty bucket and ride it down, and that would pull up the bucket with the water in it. Then you'd walk up, get in the other bucket and ride it down, and it would pull up the other bucket. We were doing something useful, but it was still fun.'

Tim reckons the river is wider and shallower than it once was. 'They surveyed the river in 1880 for the full length of the river and it was cut into trees along the bank. There was another survey crew who came back and surveyed it in 1980 … I asked 'em what the difference was and they said it had never been surveyed before. So I took 'em over and showed 'em a tree on the bank with the letters cut into it … So they looked into the records of the 1880 survey. And as a rough rule of thumb, the river from Tocumwal to Barmah is now twice as wide and half as deep.' I've heard the same elsewhere, and it makes sense. The natural river, with the annual rush of floodwaters, would have carried away silt and carved a deeper channel. The unnatural river, with its constant slow flow, is silting up. And if the flow declines further as water is rationed out, then it will silt up further, the clogging arteries of an elderly patient not getting enough exercise.

⌒

Seeking more knowledge of the river's past, I make the mistake of visiting Echuca. There are two Echucas. There is the living, breathing town, set back from the river, where life has its daily rhythms: kids go to school, parents work, council workers lean on shovels. Then there is Ye Olde Echuca, a three-dimensional scratch-and-smell recon-struction down by the river, where heritage colours have been applied so thickly and so uniformly that it's near impossible to find a primary red, blue or yellow. It's difficult to discern what's original, what's transplanted and what's fabrication. It's the Disney version of Australian history. The dock itself is impressive enough, even as a reconstruction. In its heyday it stretched for over a kilometre, the largest inland port in Australia. A 400-metre section was rebuilt in the mid-1970s, its red gum timbers towering above the placid river. Below it swarms a squadron of paddle steamers, the largest fleet in the world, their boilers still fed by wood from the Barmah forest. There's PS *Canberra*, built in 1912. A sign declares it 'The Steamboat that Heralded Today's Riverboat Trade', even though that trade was well into its decline by the time the *Canberra* was built. Another sign spells out what 'Today's Riverboat Trade' consists of: 'Weddings, Parties & Special Events'. There's PS *Emmylou*, built in 1980. Its sign declares: 'A Relaxing Journey Back in Time. Captain's Commentary'. There are seven paddle steamers in all, including a couple from the glory days: PS *Adelaide* (1866) and PS *Hero* (1874).

The wharf and boats are inoffensive, as is the tree atop the bank with flood levels marked on it: 1973 and 1974 are above my head, 1870 another 2 metres higher. I can't see the sign for 1956; someone must have souvenired it. But every good theme park has a shop, and Echuca has plenty. Behind the dock is a well-preserved, dirt-lined street. Not real dirt, but carefully raked gravel. There you can find Sharp's Magic Movie House, The Penny Arcade & Fudge, The Gift Shop & Museum, and an Antique Photographer. There's a black-smith and a wood turner and the Walkabout Aboriginal Art Gallery. There is, however, precious little indication of what life in a river port might actually have been like. What's left is sterilised, romanticised and commercialised. There's no horseshit on the river-washed gravel, no consumptive beggars or syphilitic working girls, no drunken police-men wading into drunken Aboriginals. Instead, there is the cloying smell of popcorn. I feel a bad case of 'ye oldes' coming on, a condition not dissimilar to gripe. I escape towards the main street, but there is no escape. There's the Old Aussie Grain Store—Coffee Parlour and The Old Port Cafe and the Good Ole Country—Gifts, Bears & Xmas Wares. My 'ye oldes' take a turn for the worse and I scurry back along the dirt street towards the car. And it's there, among the fakery, that I find one bona fide relic from the past: the Echuca Club.

The club's doors are barred, but behind the grill work the gilded lettering is clear enough: 'Members Only'. Once upon a time, every self-important country town boasted such an establishment. They were invariably named after their host town: the Echuca Club, the Mildura Club, the Albury Club; small-town imitations of the Austral-ian and Melbourne Clubs, themselves replicas of the London originals. They were strictly men only, with membership by invitation only. Here, the squattocracy could drink tumblers of whisky and soda with the doctors and the solicitors, aloof from the hoi polloi. Here, and not in such democratic irrelevancies as town halls or shire councils, many of the real decisions were made. Amid all the frippery and fakery of the Port of Echuca, only the Echuca Club remains unattractively authentic.

Fleeing Echucaworld, I head back to the forest, seeking restorative solitude. This time I note the long low levee dividing forest from farm, built all those years ago by pick, shovel and wheelbarrow. I drive to Lake Barmah, formed by the choke, and a favourite of campers and fishermen. Here I encounter a couple of knockabout Australians. First I meet Mick, a wiry old pensioner who came to Australia from Croatia decades ago. He's sitting on a low chair by the water, tending his fishing lines and nursing a beer. He tells me he's been coming to Lake Barmah from Melbourne for thirty years, and in all those years he's never caught a Murray cod. It doesn't worry him too much; like many Europeans, he's quite fond of carp. As if on cue, his wife, a largish woman buoyant in a screaming-pink terry-towelling dress, squeals with delight, and Mick races off to net the fish. We're soon joined by Trevor, a young bloke from Romney, down near Melbourne. He's another regular, and he and Mick have struck up a friendship over the years. Trevor looks as rough as guts: a Ned-Kelly beard, blue singlet, devil skull tattoos on his arms. He's wearing a Mobil cap turned a uniform grey by the steady accumulation of grease. He offers me a beer and the talk returns to cod. For all his rugged appearance, Trevor is a gentle soul. He catches Murray cod frequently from his boat but gives them a kiss and tosses them back in. He comes for the sport, not for trophies or a feed. He tells me he's steeling himself to try cod, just to see what it's like, and to share some with Mick. He's going to take the old migrant out on his boat, up into Lake Moira, so the old man can experience the thrill of catching the cod that has eluded him from the banks these past three decades.

Mick tells me he and his wife love coming to the river. 'One night we camping here. Plenty drinks. Plenty noise. The ranger come. We thought oh-oh, too much noise. But no. He say big flood coming, we have to move. Sure enough, next day, all this was flooded. We were on high ground. A big bunch of us, all together. It was all okay. Like a big party.'

161

Trevor reckons turning the forest into a national park 'is shit'. He fears the informal sense of community that has built up along the banks of the Murray over all these decades will be lost. I'm not sure what Mick thinks. His wife lets out another squeal and he heads off to net another carp before I can ask him. If the national park people do change the ways of the fishermen, I just hope they don't try and compensate for it by building another theme park dedicated to the good old days.

⌒

The group of angry men are waiting for me outside the cafe in the little town of Barmah, where the forest ends and the river returns to the steep banks of normalcy beyond the choke. They're not angry with me, they're angry with the Victorian Government. They're angry at it for stopping the grazing, for stopping the logging, and for turning the forest into a national park. The angry men stand in a loose gaggle around a collection of four-wheel drives and work trucks, each one emblazoned with bumper stickers in red and white: 'Rivers & Red Gums for all and forever!' They give me one for the back of the Hyundai; I tell them I'll put it on later. Here I find the owner of the Barmah sawmill, David Swan, and his 83-year-old father, Jack. David's brother, Kevin, who owns another mill at nearby Picola, is there with his son, Trent. The old horseman Gerry Moor, friend of Tim Mannion, is here. So is Leigh Wright, a self-published author who has recorded all the old forest stories. And so too is Peter Newman, chairman of the craftily named 'The Rivers and Red Gum Environmental Alliance' and secretary of 'The Barmah Forest Preservation League'. Peter is a forceful advocate, describing the government's actions as lunacy. It's just a month after the Black Saturday bushfires have devastated Victoria and he argues that without grazing and logging, the forest will become a tinderbox, especially now that flooding has become so rare. 'The whole lot could go', he warns. 'The whole bloody lot.' It strikes me as an opportunistic argument, but that's not to say it has no validity.

I'm not alone. A documentary maker named Gerhard has come to do some filming for German television. We have a quick confab and decide to take the men into the forest; Gerhard needs to have them on location for his camera. So sawmiller Kevin Swan jumps in the front seat of the Hyundai and we head off after the others. A light misty rain starts to fall as we leave the bitumen and enter the forest. I'm not too sure how my city car is going to handle the sticky mud of the flood plain if the rain gets any heavier, but Kevin thinks we should be okay. As we drive, he tells me his story, his heritage. He's fifty-three years old, a fourth-generation sawmiller, his family having worked the forest since 1853. He's spent his entire life in and about the forest. He has no illusions what the declaration of the national park will mean for him. 'I'm out of business', he says. 'Our licence is for this management area. And there will only be approximately 10 per cent of the red gum forest left available for logging. And that's not in this area, that's 150 kilometres away.' Kevin says that the government has identified fifty-six timber workers who will lose their jobs, but that compensation is capped at $80 000 each. He says that will go nowhere close to recompensing him for his plant and equipment. And with the industry shutting down, who would want to buy it from him? 'I'm too young to consider retirement, I'm too old to be considered as a valuable employee', he says matter-of-factly.

By now we've penetrated deep into the forest. In the drizzle it looks somehow healthier, as if even the suggestion of moisture is enough to buoy the trees. Kevin points out a dark line about a metre up their trunks, a line he says was left by the last big flood. We catch up to where the others have stopped. Gerhard has corralled them by the foot of a forest giant while his cameraman dances about capturing different angles. Kevin climbs out and swaps places with Trent, a fifth-generation forester. Trent is twenty-three. He tells me he left school at fifteen to become a sleeper cutter. It's all he's ever wanted to do.

'My employment opportunities around here are very minimal with the drought. So to work, which I'll have to do because I've got a mortgage and everything like everyone else, I'll just have to move away and get a job. All I've ever done is cut sleepers and work in a sawmill. So I'll have to move away to work in another sawmill somewhere else … It will be pretty difficult for Dad. He'll be the same as me, maybe worse off. I'm a bit worried about him. He's been in the forest for thirty-four years officially, ever since he went to work with his father. That's what I would have liked to have done. We're a father–son team, the last father–son team cutting in the Barmah forest, and I would have liked to have taken over the family business cutting sleepers in the Barmah forest.'

The rain picks up a little in intensity and Trent looks glumly out into the forest that will soon be denied him. It's not just the drought. The global financial crisis is threatening to unleash its full fury on Australia. Twenty-three years old, a mortgage, left school at fifteen, no experience beyond cutting railway sleepers. Terrific.

I decide to get out of the forest while I still can. Trent climbs out and Gerry Moor, the old bushman, climbs in. I inch the car about and start following my tyre marks back. Gerry knows the forest like the back of his hand; he's descended from the first white settler in the district. After thirty-five years away training racehorses he's come back to the district. He decides to take me back a different way; he wants to show me another part of the forest. I'm a bit wary about this, but I figure if Gerry doesn't know his way about, then no-one will. About two minutes after I follow his directions, we're bogged. The rain puts on a celebratory spurt and I refresh Gerry's knowledge of the bullocky vernacular. He climbs out, examines the situation, and tells me to floor it, to spin the wheels. It's the direct opposite of everything I've ever learnt about avoiding getting bogged, that spinning the wheels will only bog you deeper. I remind him of a few more ripe expressions of yesteryear, but he just laughs. When I climb out I see what he means. We're not bogged, not in the traditional sense. The front-wheel drive hasn't dug itself into the ground; it still sits on top. Instead, the mud is so sticky it's clinging to the tyres and has built up inside the wheel wells, locking the back wheels tight. The mud on the track is only a few centimetres thick; underneath, the ground is dry. I reverse a little to displace some of it. 'Break through the top!' laughs Gerry. 'Spin the wheels!' Who am I to question an old bushy's wisdom? I plant my foot, the wheels spin, and Gerry is covered in mud from head to foot. I make about 10 metres before the back wheels lock again. It takes another forty minutes of this to get free: reversing, scraping mud, moving forward. Two steps forward, one step back. Gerry, to his credit, keeps laughing.

⌒

Many weeks later I walk from Woolshed Creek out onto the Chowilla flood plains in South Australia, not far from the Victorian border. Like Barmah-Millewa, Chowilla is one of the six icon sites identified by the Murray-Darling Basin Commission and listed as a wetland of national significance. Like Barmah-Millewa, it's an internationally recognised Ramsar wetland. Like Barmah-Millewa, it's protected by migratory bird agreements with China and Japan. But Chowilla is dying. I walk through a landscape of dead and dying trees. The grey silt of the flood plain is bare and drifting in the wind, the undergrowth gone. It's almost like a fire has been through. The ash-like dust covers everything: it's spread up the trunks of the trees and out along the branches to what little foliage is left. Chowilla hasn't had a real flood since 1993. The scarce water that's available has had to be rationed out, sent into a few select areas. But where I walk, the only water has been rain, and in the worst drought in a century, there's been precious little of that out here in the semi-desert. Even by the creek, where there is water, the trees have been losing branches. Later I learn the water is intermittent, that it has been let into the creek just recently, well into autumn. If Barmah is sick, then large parts of Chowilla are terminal. There's no choke here to subvert the designated flows; no water escapes unless it has been debated, countersigned and rationed. Somewhere, in some office, the authorities are making difficult choices, deciding which parts to preserve, and which to let go. The great river red gum forests can survive drought and they can survive the artificial river, but they can't survive both. The terrible greyness lies everywhere across the land.

7

WAKOOL

The Wakool River: 9 per cent allocations and gallows humour

I'm driving flat chat as I cross Green Gully, a nondescript depression in the wide flat plain of western New South Wales. It's not green at all, not even a gully, just a long shallow scarring of the earth, with a spattering of low grey bushes struggling to distinguish it from the surrounding grassland. Once, water ran freely here, when Green Gully was the bed of the Murray River, back before the Cadell Tilt sent the river sliding off to the north and south. Several thousand years later, there's precious little to suggest the verdure that once tracked the river through here. It's a reminder, if one is needed, of how ephemeral water can be west of the mountains. Even the continent's most vaunted river can dry up, vanish, never to return.

But I don't stop to contemplate Green Gully or its portents; I'm in a hurry. Indeed, the only reason I'm passing this way, pounding the faithful old station wagon across the corrugated surface of an unmade road, is because I'm running late. So I plough on, trailing a plume of dust, a few more parched grains to fall in Green Gully, a few more specks to assist in the millennia-old project of erasing it and its uncomfortable message from the landscape. But an hour later, back on the bitumen, I can't help but stop when I get to the Wakool River. It isn't there either. The bed is almost as dry as Green Gully, just a puddle here and there. Protruding from the cracked mid-stream mud is a blue beach umbrella and an old chair, as if set up for a picnic. Gallows humour: a sure sign that things have taken a turn for the worse. I continue on my way, but it's difficult to push the accelerator home when you have some idea of what may be waiting. Nevertheless, I arrive in Wakool soon enough. 'Population 350, Elevation 80 m' says the obligatory sign, its surface peppered by gunshot pellets. The sign is an old one—the 2001 Census put the population at 225, and the 2006 Census put it at 213. And that was back in the good times, before the drought really bit.

I check into the pub, a modern place, big and welcoming. I'd been worried that there might not be any rooms left, what with the town's annual show just three days away. Anne, the publican, a big woman with dyed red hair, offers a dry laugh: so far I'm the only guest. She leads me out the back to an unpretentious room with twin beds; $25 a night, with the bathroom and toilet down the hall. Perfect. Back in the bar, Anne gives me directions to the services club, where the show meeting is being held. It's on the next block, 50 metres along what serves as Wakool's main street. As I go, Anne says she'll leave some dinner behind the bar for when I return.

Tuesday night must be meeting night in Wakool. Who could imagine so much communal activity in such a small place? The P&C is meeting at the school, the Land and Management Committee is in one room at the club, and the show committee is in another. It's the committee's final meeting, a chance to iron out any last-minute glitches. I arrive in time to meet the members before they get down to business. Mick the president is there, a sardonic haystack of a man cradling a large bottle of Diet Coke. Andrea the secretary is there, one bright eye on the minutes and the other on her 7-year-old son, Sam. He's bouncing around the periphery of the meeting, making rocket noises with his mouth. Gail and Andrew, both in their fifties, have come in from their dairy farm, and Max is there too, crow-feet eyes suggesting lots of sun and a ready laugh. Old George is there as well, with memories of more shows than he cares to remember, his experience rivalled only by 74-year-old June, the pillar of Wakool. They make me welcome and, a little self-consciously, start addressing their agenda. But I'm soon forgotten as the imperatives of organisation assert themselves.

The shire council has mowed the lawns and cleaned up the oval. The footy club is happy for its clubhouse to be used now that the old show pavilion, riddled with white ants, has been demolished. Friday night will be the show's big night, with Saturday reserved for the horse people, although there's been a cock-up with the ribbons for the equestrian events—Andrea is getting them express-posted from Melbourne to Barham, 35 kilometres away on the Murray. Sideshow alley will go on the far side of the oval, the band will go in a marquee near one end of the clubhouse, and the truck with the Miss Showgirl competition will be at the other end. Sideshow alley is looking promising; this year's show no longer clashes with a nearby rodeo. There'll be a couple of jumping castles, a spray-on tattooist, dodgems and the octopus. 'Not the octopus! I hate that ride!' interjects Sam. Max says that the band, the hottest thing to come out of Deniliquin in years, is costing 'a motza', but that he's hoping to entice a Deni bus company to bring people the 60 kilometres from the regional centre. The ice-cream man will be back, having had a good day at last year's sheep races. The fireworks guys are also returning. The announcer from last year is happy to front up again in return for free entry, a couple of beers and a feed. June says she'll be making sandwiches for the judges and Andrea confirms she can bring in a whipper snipper to clean up around the sheds. Max says that despite tough times, all of last year's sponsors are back again.

There's a debate about the best price for raffle tickets: $2 each or three for $5 is the verdict. The donated prizes have started coming in: vouchers for meat trays, a case of rosé from the pub, some coupons from the shops in Barham. There's some talk about avoiding last year's problems. Max says the fireworks were too close to the marquee, which came under ember attack. Luckily the volunteer fire brigade had just finished their fire-fighting demonstration and were on hand to keep an eye on it. And the sideshow people hadn't been impressed by a 3 a.m. burnout and doughnut demonstration by a bunch of local youths right across from where they were camping. When they had emerged to remonstrate, the young rev-heads had mooned them. But the ringleader, a Greek girl called Nikki, has moved out of the district, and there's confidence the sideshow people will get a reasonable night's sleep. The police, aware of the trouble at the recent Deni show, will call by from Barham.

The meeting winds up. June reminds me my dinner is waiting for me behind the bar at the pub.

⌁

From the air, Wakool looks the same as it does from the ground: nothing much. It's six or seven blocks, a railway line and a couple of irrigation canals, with the tin roofs of the pub, the club and the school surrounded by those of a few dozen houses. The school oval and the footy ground look like green circles painted onto a brown hessian canvas. There's one dairy farm close to town, where the bore-fed fields glow iridescent green. Everything else is faded brown, stretching to the horizon. Allan Adams banks his half-century-old Cessna more steeply and we trace a tighter circle over the little town. I can see the full irrigation canal and the empty river. Out along the Barham road, we fly low over the dry bed of the Wakool and I spot the blue beach umbrella and chair, still mocking passing traffic from the bottom of the empty river.

Wakool lies about 35 kilometres north-west of the Murray, up where the river has split into its mid-stream delta, about halfway between Albury and Mildura. Water backed up behind the Barmah Choke is pushed up the Edwards River, where it splits again into the Wakool. That's supplemented by water travelling down the Mulwala canal from the weir at Yarrawonga. The land is relentlessly flat, perfect in many ways for irrigation, especially back in the days before bull-dozers and laser levelling. The Edwards–Wakool system is entirely gravity-fed. There was a time when engineers from around the world would visit to pay homage to its design. But if drought has hit irri-gators hard, nowhere have they been hit harder than at Wakool. The foreign engineers, along with everything else, have dried up. Two years ago, irrigation allocations at Wakool were reduced to zero. Nothing. No water at all for irrigators, no matter what their so-called entitle-ments. The same last year. This year's allocation, set at 9 per cent, is too little to grow anything. It would be an insult, except the irrigators can on-sell their 9 per cent, raising a little cash from their only saleable commodity.

Allan and I fly over a parched landscape. Some fields are not just brown and dead, they're completely barren, devoid of anything except bare dirt. Some display the straight-line levees of laser-levelled rice fields, others the contour-hugging banks of the original fields. We fly over a property where someone has planted saltbush. 'Mad idea', says Allan through the intercom. 'He's long gone.' We come in low over the huge grey-metal silos and packing sheds of the Burraboi Rice Depot, the largest rice silo complex in the Southern Hemisphere. But there's no sign of activity: no trucks, no parked cars, no ant-like figures. Its 55 000-tonne capacity lies empty, its sheds mothballed for want of rice. We fly on, above an experimental aquaculture farm similarly mothballed for want of water or capital or hope: Allan isn't quite sure which. We cross the evaporation pans of the salt-interception scheme. No activity here either, but the artificial saltpans, shimmering in the sun, don't lack for raw material. Then we pass above a solitary rice farm that has somehow garnered enough water for a crop. The fields, watered by some unknown combination of bore water, saved water and purchased water, shine impossibly green. But the contrast only serves to emphasise the pervasive sterility of the surrounding fields.

It should be depressing, but it's not, thanks to 64-year-old Allan, whose humour is as dry as his land. Allan reckons rice farmers are being demonised by the know-it-alls in Melbourne, the same know-it-alls who are commandeering irrigation water 'to flush the capital's toilets'. Allan's father was one of the first six farmers to grow rice in the district, back in 1948, following on from a huge wartime crop grown by Italian prisoners of war. Allan once had several planes but now he's down to just one, his beloved 1960-model C180 Cessna. He doesn't blame the drought; rather, this is the result of what he obliquely calls his 'family law debacle'. He can't quite bring himself to refer to his former wife by name; she, too, is simply the 'family law debacle'. I ask what happened. The answer is vague—financial strains imposed by the drought seem to have something to do with it. Some things are better left unsaid.

⌒

Andrew and Gail Tully sit in their neat farmhouse and recount the story of their prize-winning dairy. Andrew is balding but his remaining hair is still dark. Gail's a friendly woman, with long streaked hair and a ready laugh. She's made scones and jam. Country hospitality. They live just outside Wakool on the 300-hectare soldier-settler block drawn by Andrew's grandfather in a 1949 lottery. The block sits snugly against the main southern irrigation canal, a canal spitefully full of water despite the paltry allocations available. Until recently the farm was highly successful, winning awards for quality year after year. It was so successful that by the start of the decade, the Tully's herd had grown to 300 milking cows, plus another 150 younger animals. The farm supported three families: the Tully's and those of their two full-time employees. Milking the herd was taking up to eight hours a day, so the couple took the next logical step, investing half a million dollars in a state-of-the-art rotary dairy. When they started using it in February 2002, they couldn't have been happier. Decades of hard work were coming to fruition. Andrew says that the intention was not to expand the herd any further, but to concentrate on improving quality. Then, in the summer of 2003–04, their water was cut to just 8 per cent of their nominal allocation, the first time such a drastic cut had been imposed. Even during the crippling drought of 1982, their allocation had only been cut back to 80 per cent. One farm in the district closed but the rest toughed it out, buying either feed or water or both.

'We sold all our … shares and bought temporary water, thinking this is just a one-off. We extended the overdraft. And we never really recovered from there', says Andrew. Over the next two years, 2004–06, the annual allocations were 46 per cent and then 54 per cent. 'We just couldn't catch up. We were scraping. Things like normal maintenance operations just weren't happening. We weren't putting on as much fertiliser as we should have. The farm kept operating, but if we'd kept on going like that we probably wouldn't have survived, because we weren't putting money back into the business.'

Worse was to follow. In 2006–07, the allocation was zero. In 2007–08 it was the same, before this year's desultory 9 per cent, while the drought ensured that rainfall remained significantly below the historical average of about 30 centimetres.

'In 2003 and 2004 there were nineteen dairy farms operating in this Wakool area', says Andrew. 'And there were six rotary dairies like ours. And there were probably another five or six big herringbone structures that people had put a lot of money into. A lot of those farms had two or three families, like ours. The farm next door was about 600 cows. He went broke in the first year because he had only just launched into a larger dairy. The other rotary at Tullakool, they couldn't keep going. They had to sell their cows, sell their water. And in that time, during those six years, people have just dropped off. They just couldn't afford to keep going. There were nineteen dairies, now there are three.'

When the first zero allocation hit, the Tullys, already in debt, borrowed another $200 000 to get through that year. They were fighting to keep their herd, built up over sixty years. The family had never bought cows; they had bred their own, bringing in bulls and, later, artificial insemination to improve the bloodlines. But when the second zero allocation hit, Andrew estimated they would have had to borrow another $400 000 to get through another year. There was no alternative. The Tullys started selling off the herd.

Dairy farmers are different from other rural producers. They get to know each and every cow: their personalities and their quirks. They're there when they're born, raise them from calves, milk them every day. They get to know which cows are bullies, which ones are skittish, and which ones are smart enough to get an extra feed when no-one is looking. Many dairy farmers give their cows names; the Tullys used numbers, but to this day they talk about them as individuals, not commodities. 'We knew every cow and every heifer ... We had an emotional attachment', says Gail.

'It was a disaster, an absolute disaster', adds Andrew. 'There were heaps and heaps of sales going on.'

'A lot of the old girls were sent to Shepparton', says Gail. 'Plus the young heifers that were going to calve. At the end of the sale ... there were only two buyers left bidding: the two butchers. And I was thinking, these beautiful heifers that were due to drop a calf in a month or two, and the butchers were bidding on them. It was just heartbreaking. We left then.'

'I still wake at night sometimes, wondering about old 66', says Andrew.

By selling the milking herd, the Tullys had hoped to endure the drought. They kept the 150 young animals, about 100 of which are now pregnant and will calf in September. Once they give birth, their milk will come on and the Tullys will be able to reactivate their half-a-million-dollar dairy—the dairy that has been accumulating dust and debt for the last two years. But they need water to grow feed. The allocations for the 2009–10 irrigation season are due to be announced from August. I consider what I saw at Dartmouth and Hume, the dams at near record lows. The prospects don't look good. To make matters worse, the global financial crisis has hit the demand for dairy products in Asia; the wholesale price of milk has collapsed.

'These are the last of the girls. These are the babies of the girls we sold back in September 2007', says Gail. 'They're in maturity now. Once these heifers are gone, we've got no dairy stock left. We're coming to the crunch time … They'll be calving in September. So now we'll have to make a decision: will we sell them, because we've got 9 per cent water and no pasture? Or do we buy in feed so we can milk them? I think we would probably have to sell them.'

Before I leave, the Tullys take me to see the rotary dairy, standing forlorn at the centre of the farm. The wind is picking up and is keening through the wire fences. The day is overcast, with the hint of a chill in the air, the first precursor of winter. With residual pride, Andrew describes to me how the dairy works, how efficient it is, how gentle it is on the cows. And it's evident, even to a city boy like me, how well it has been constructed. And how expensive it would be to remove it for sale. If there were a market for it. Andrew is fifty-six, Gail fifty-two. If they sell the last of the herd, they will be too old to restart, even if the water returns. For the Tullys, it's the last throw of the dice. And they know exactly what the odds are.

⌣⟶

John Licato's rice farm sits on the dense black soil of the Wakool flood plain. He hasn't planted rice for three years. Rice farmers are often criticised for their profligate use of water, but John has a different perspective: he says they only put in a crop when they have water. Before the current hiatus, the farm had grown a rice crop every year since the late 1940s. There's not the same heartbreak here as at the Tullys, but there are all the ongoing costs, the debts to be serviced, the water supply fees to be paid. For even if farmers receive a zero allocation, they are still expected to contribute to the upkeep of the irrigation system. Selling their 9 per cent allocations on the temporary water market will leave Wakool's farmers very little change after they have paid these fixed costs.

John is a difficult person to interview. Not because he isn't forthcoming; just the opposite. He jumps from farming to water to politics to history to agriculture to finance to family and round again with such rapidity that it's hard to keep up. He blames his Italian heritage.

With a zero allocation last year, John gambled on putting in a dry-land winter crop. It cost about $50000 to sow. But it didn't rain. The harvest yielded a truckload of barley and half a truckload of wheat. Now he's deciding whether it's worth punting on another winter crop, what the locals once dismissed as 'Mallee farming', back when they were cushioned by the assurance of irrigation. 'This year is the crunch time', says John. 'I'm relying on this winter crop. If you have a look at my farm, it's fallow from one end to the other. There's not a blade of grass on it. So I'll crop again this winter, but whether I'll go the big lick again, I don't know. Because I'll have to buy fertiliser and, for the first time, seed. Plus fuel of course.'

John reckons Wakool has been singled out for obliteration. He's not alone. There's a widespread belief that the faceless men and women of Canberra have drawn lines on maps, deciding who shall prosper and who shall perish. So last year, the region's 230 irrigators gathered together and offered to sell their permanent water to the government as part of its much-spruiked buyback scheme. The farmers said they'd close down the entire irrigation system provided the government paid them enough for their water. On the table was 330 billion litres. The attraction for the government was that, as well as the water actually used on the farms, it would gain all the water wasted in conveyancing. In many ways the proposal was a stunt, designed to call the government's bluff, as about 30 per cent of the irrigators didn't really want to sell. John said the government wasn't interested; the price the irrigators were demanding was too high.

John says his grandfather thought he had found the promised land when he came to the district from Italy in 1923. But John, who has two sons, one of them keen on farming, doesn't want the farm to pass to a fourth generation—the promise has faded. 'Not if I have anything to do with it. There's no way known I'd let him come back into farming ... The way it's structured now, there is no future in broadacre irrigated agriculture.'

And that's exactly what Wakool is: broadacre irrigated agriculture. Rice and dairy. Back in its heyday when the foreign engineers would come to marvel, the gravity-fed canals of the district, stretching back through the Mulwala Canal to Yarrawonga weir, were seen as efficient and low-cost. Water was cheap and plentiful, and there were no expensive pumps or pipelines. Flood-irrigating dairy pasture or inundating rice paddies was a profitable business. But now water is expensive and scarce, and the evaporation and seepage in the canal systems are not so easily ignored. Nor are the opportunity costs involved in irrigating large areas for seasonal crops or dairy. The 160-kilometre-long Mulwala canal doesn't appear quite as marvellous as it once did, and so the engineers have moved on, interested now in pressurised pipes and computer-monitored sub-surface micro-irrigation. John Licato is right: the economics of large-scale broadacre irrigation aren't what you'd want to bequeath to the next generation.

⁓

I take a walk around Wakool. The main street is a one-sided affair, lined by buildings on its eastern flank, but with none opposite. Across the road from the pub there's a small park, then the main road, then vacant land, then the railway line. In the park, a sign recounts the major events in the town's history: the arrival of the first passenger train in 1926; the opening of the picture theatre in 1954; the start of the fisheries research project in 1998. The dates of the last passenger train, the final picture show and the mothballing of the fishery project aren't recorded. Next to the park there's a traffic circle known locally as 'the bullring', where the war memorial rests among respectfully trimmed greenery. The memorial is a low concrete plinth, painted white. It supports a modest flagpole and plaques listing the forty who served and the one who died. Wakool didn't exist back during World War I, when so many country lads died on the Western Front. Instead, the town is part of that final postwar push into the bush, sited on an anabranch of the inland sea, settled even as the motorcar and mechanised farming began the long slow depopulation of rural Australia.

Along from the pub are a couple of old shopfronts that once housed the local Elders outlet. Here I meet Charlie, a recent arrival, who is working on putting in a new downpipe. In the window of the old shop, I can see a teddy bear. Charlie tells me he bought the place for $40 000. He reckons he can do it up and sell it for $100 000. He points across the railway line, says there's a place over there going for $60 000, tells me to snap it up. He says the locals think he's daft.

I suspect they're right. I ask him if he's come here with his family. No. His wife is up in Brisbane; they're splitting up. Next door to the old Elders store is the remains of a supermarket that closed decades ago, but the outline of the words 'Wakool $upa Valu Supermarket' is still legible where the signage was removed; Charlie says the building is condemned. A derelict petrol bowser sits rusting beside the road, bearing the remains of an old BP insignia. I check to see the final price of petrol, but the bowser has been gutted and its innards removed.

A little further along, past a couple of houses, the local store-cum-post office is open, housed under the semi-circular roof of a Nissen hut. Signs advertising the *Herald Sun* and the *Weekly Times* betray a greater affinity for Melbourne than Sydney. It's like that throughout the Riverina: Victorian papers, Victorian beer, Victorian football. The signs share the space above a cursory awning with a hand-painted Santa Claus who is missing his cap and beginning to fade after a summer in the sun. A few drops of rain unexpectedly start to fall, so I go inside and order a coffee.

'Cappuccino or flat white?' asks the man behind the counter. I order a flat white, which turns out to be the local name for instant coffee with a splash of milk. I make no complaint. 'Latte drinker' is one of the great modern pejoratives of the bush. Instead I get talking to Annette, a largish woman who owns the shop with her sister, Sandra. Annette's busy making a plate of sandwiches for some pensioners gathering in the memorial hall. The two sisters moved up here from near Mount Macedon to get away from the cold weather. She says they like it in Wakool, that the locals have welcomed them into the community. The sisters drive to Barham three times a week to pick up the papers, milk and bread, a 70-kilometre round trip. But business has gone down by 50 per cent in the last year and they've gone into debt to keep going. The sporting teams that used to come in after training for takeaway have stopped coming, either because they no longer exist or because individuals are watching their pennies. Annette feels the town is becoming stretched and says that this year's Australia Day breakfast saw only six families turn up. Sandra thinks there may be an element of 'duty fatigue' setting in but people are still up for a bit of fun. She signed up along with seven other women for a weekly bowls night at the local club and so far it's turned out to be a great success. She says there's still strong support for the annual fishing classic, the sheep races and, of course, the Wakool Aussie Rules team. Annette says she hopes the show goes well, although there are predictions of rain. If so, it would be the first rain this year.

I resume my walk around the block, turning left into La Perouse Street. The town's half a dozen or so streets are named after great seafaring explorers: Tasman, Dampier and Cook, Flinders, De Quiros and Bass. Such water-laden names for such a water-starved town. A few houses along La Perouse Street, I reach the Wakool Memorial Hall, just as Annette is dropping off her sandwiches. I call in to say hello, knowing June will be there. I find five elderly women and one man, sitting round a table in the kitchen playing bingo. The sandwiches are the signal to finish up, to have a cup of tea and a bite to eat. I get talking with Jack Main, a former rabbit trapper, shearer and farmer, now eighty-five years old. He has the most arresting eyes, bleary brown irises rimmed by a halo of blue, as if after too many years under the pale Riverina skies, some of their colour has leached down into his eyes.

Jack's an old fisherman. He reckons that despite the parlous state of the local rivers, the Murray cod have never been better. He says the cod in the Wakool are different to those in the Edwards and the Murray: shorter and thicker. He tells me he once caught a 20-pound cod, and when he opened it up he found thirteen mice in its stomach; he says it still makes him think. He reckons climate change is a lot of nonsense, that even the recent heatwave, with daily temperatures of over 40 degrees for a fortnight, isn't anything compared to 1939, 'the hottest I've ever been in my life'. He says in 1939, 1940 and 1941, the Murray was so choked with slime it was impossible to fish. Besides, he tells me, the present drought is about to break; Lake Eyre is filling and whenever that happens, relief can't be far away. It's a popular theory in the bush.

While Jack and I talk, the women clean up and a couple drift away, the social whirl over for another week. I chat with June, a community powerhouse. She has eleven children and twenty-nine grandchildren. She's on the show committee, the memorial hall committee, the welcome centre committee and in the Wakool Action Group. She does voluntary work for the church, helps with the annual fishing competition, and is a life member of both the P&C and the Wakool football club. 'I run the canteen. I've told them this is my last year, but they don't think I'm serious. And I'm probably not, but don't tell them', she says. Last year Wakool didn't win a game, but the football team remains the heart of the town. June once had five sons in Wakool's premiership side and they were featured on the *Midday Show*. Now only Ginger plays—he's forty-five but keeps going to help make up the numbers. Ginger, a shearer, is married to Anne's daughter, Chelsea, and so June also helps out at the pub, looking after the couple's young boys as well as doing washing and ironing and cleaning the rooms. 'I don't get paid for any of it but I enjoy doing it. And while I'm doing it, I'm not going to be crippled up with arthritis sitting around and feeling sorry for myself', she says.

Nevertheless, June has been forced to give up a few activities. She joined the local chapter of the Country Women's Association to keep its numbers up, but it collapsed anyway two years ago, sixty years after it was established. After thirty-three years, she no longer cleans the local school. And she no longer works as the Wakool correspondent for the Barham newspaper, *The Bridge*. 'I gave that away. I was always on the phone, trying to find out stuff. We've got a neighbour across the road who doesn't have a kind word to say about anybody. I give it back to him, double. Anyway, one day he took his wife out in a boat. She's a very big lady. She's lovely, but big. Anyway, he tipped her out of the boat. And I put it in the paper. Didn't mention any names or anything, but he still went for me. "You slanderous woman!" he yelled at me. "You slanderous woman you!"'

June leads me out of the kitchen, through the hall proper, where badminton courts are outlined on the polished wooden floorboards, and towards the stage. She wants to show me a painted backdrop behind the curtain depicting the old Tullakool homestead, but it's been covered over by black cloth. She explains that it was painted for a debutante ball five years ago, Wakool's last deb ball. 'I used to train debutantes. Every year there would be a deb ball, and I would teach them to dance. And every year we would have a different theme. One year we had a water theme; we had waterwheels. A lady up the road painted a Murray River backdrop. It had ducks on it. It was really a beautiful thing.'

From the stage, June describes the balls to me. She speaks in the present tense, as if they still happen, and her eyes move around the room as she talks, as if she is watching the young girls in their gowns and their beaus in black tie.

'The girls come in through that door up there ... and they come down here and do their curtsy. And the boys wait down here. They all bow to the mayor and to whoever they need to bow to. Then one goes this way and one that way until they make a circle. And they do different dances they've been trained to do. It used to be a really big night out, but it doesn't happen anymore. There's not enough girls. We had it on a Friday night once, and so on the Saturday night we still had all the decorations and the band and everything still here ... we had a ball for older debs. I'd never made my debut 'cause I never waited long enough to wear a long dress, although I could dance since I was eleven. It was a wonderful night. Talk about fun. This hall has seen some really good times. My husband and I had our fiftieth wedding anniversary here. He's passed away now, three years ago. Really good times.'

June adds, 'They used to have a rice ball here every year, but whether it will eventuate this year or not, I don't know. That's pretty big. They come from all over. They dress it up with rice and things. One year they had all pictures of machinery around the walls. It really does up beautifully. But I don't know if you can have a rice ball if there isn't any rice.'

It's not always easy to talk with women in a country town, not if you're a stranger, not if they know you're writing a book. You need to get them by themselves. If there are men around, their husbands or others, the women are inclined to defer to them, busying themselves making tea and small work while the men discuss rain and markets and whichever politician is particularly loathed this week. But in a time of drought, it's the wives who go out to work, travelling to the big towns, to jobs at schools and hospitals and nursing homes and shops and cafes and dentist surgeries and banks, who bring home meagre pay packets while the men stay on the farms and fret. And it's the women who play a large part in keeping the community together, organising events and rallying to help those who have hit hard times. It's part of the old ethos of the bush, out where government services don't always reach, where people are self-reliant but know that at times of crisis they have the support of their neighbours. It's how country people often distinguish themselves from city dwellers, how they justify staying where they are. Time and again, people I meet acknowledge that moving to the city or a big town would make financial sense, but that they value the community and lifestyle of small-town Australia.

I think of this as June locks up the hall, and I continue on my walk. I turn left again, onto Bass Street, a narrow strip of bitumen laid down the middle of a wide dirt road. The Murray Irrigation depot takes up a whole block. Three dozen Dethridge wheels stand in bright array, galvanised and ready for action. But the wheels are unlikely ever to be deployed; the government is funding the conversion of the irrigation districts to electronic gates. There will be no need for the old wheels, nor for those who travel the canals reading their metres. Already, Murray Irrigation, the town's second largest employer, has been laying off staff. I pass the spick-and-span preschool and turn left into Phillip Street, past the primary school, an oasis of government funding in the drought of rural hardship. On the other side of the road, the old CWA house lies decrepit and collapsing, paint peeling from its fibro and weatherboard skin; there's asbestos inside and the building was abandoned even before the CWA itself collapsed. I turn left again, back into Flinders Street, past the Uniting Church, its cross askew and a sign reading 'Worship—most Sundays'; past the services club; past the old post office, its postboxes rusted shut. The post office is now used as a library for three hours on Wednesday afternoon. And that, apart from a few houses, the footy ground, the bowls green and a petrol outlet out on the Deniliquin Road, is about all there is to Wakool. I'm back at the pub.

I walk into the hotel's front bar and order a beer. There's only Anne's daughter Chelsea, behind the bar, and her two boys, 4-year-old Jake and 14-month-old Mitchell. Chelsea has long strawberry-blonde hair and a toothsome smile. Three years ago Chelsea and her mum were commuting to Barham to work in the services club. But then Anne's other daughter bought the pub and installed her as the proprietor. Despite the loyalty of the regulars, the women are now selling half the beer they did three years ago.

'The first year we had fifty in the footy tipping comp', says Chelsea. 'Last year there were thirty. When we looked at the board this year before we rubbed it off, we counted that twelve of those thirty have left. We'll still get about thirty this year, but they'll be different people.' That's not as bad as it sounds. Many of those that left were workers from the feedlot 20 kilometres north-west of town. The feedlot started laying off workers last year, but then it was sold to a multinational with deep pockets and it started rehiring. With the original workers already gone, water-starved farmers took the jobs. At least twelve are working there, seeking shelter from the drought. It's helped to keep the town going: farmers work at the feedlot and their wives work in Barham and other river towns. The hope is that once the water comes back, the farmers will go back to their properties and new workers will pick up the slack at the feedlot and boost the population. It's a good theory, something to cling to.

An old farmer comes into the bar. The talk is immediately of rain. The afternoon's few sporadic drops have lifted the farmer's spirits and he's come to town to share the news. He tells me he's had 4 millimetres of rain during the day. I say it doesn't sound like much. He says it takes the year's total, two and a half months in, to 5 millimetres. He'll take whatever he can get, thank you very much. I buy him a beer.

⌣

Later on, I duck back into the pub to buy a bottle of wine. I've been invited round for dinner by Andrea, the secretary of the show committee, and Merv, her husband. 'Where are you off to?' asks Chelsea. 'Andrea and Merv's? Oh, take this one then. Andrea likes this one.' Chelsea hands over a bottle of red. I wonder if there is anything the townspeople don't know about each other.

Andrea and Merv live round the corner in La Perouse Street, opposite the memorial hall. The rain is now nothing but a memory and we sit out the back while Merv fires up the barbecue: sausages and local steak. I'd wanted to talk to Andrea about the town's women, but she defers to Merv and takes over the barbecuing as we talk water licences and canal management. He's a big burly bloke who loves to talk. But he tells me he doesn't get out that much anymore. That's because he works for Murray Irrigation. Some irrigators can't resist having a go at him, even though they know their lack of water isn't his fault. Maybe they resent Merv's reliable income, paid from their standing water charges, when they have little or none. And Merv knows how tough some of those farmers are doing it, how close some of them must be to bankruptcy, because he knows which ones struggle to pay their water bills. He mentions no names, but says some haven't paid for two or three years. No wonder he prefers to stay at home with the family.

Sam is bouncing around excitedly, showing me his rock collection. Merv dotes on him. He helps Sam fire up his Peewee 50 motorbike and Sam guns it round the backyard, doing circuits and chasing the chooks while Merv and I talk. At fifty-six, this is Merv's second marriage. His first broke up after twenty-four years. He tells me he can't believe his luck, that he's blessed to have Andrea and Sam. He repeats this every twenty minutes or so, as if verbally pinching himself to check it's not a dream.

After dinner, Andrea heads off to clean the school. She took over the job when June gave it up after three decades. It's only fair; Merv's mum used to do it before June. Merv tells me about his own job. He's negotiating with landholders to decommission canals. He tells me about the 54 kilometres of canals branching out along two spurs towards Moulamein, supporting twenty-two irrigators. Twenty-one of them want to sell to the government if the price is right. There's one holdout. That's a problem for Murray Irrigation, because it would still be obliged to supply him, even though it would take more water to get his entitlement to his farm than he would use. The company isn't allowed to pressure the holdout to sell, but Merv thinks his neighbours will convince him. If they're successful, they'll be able to extract a higher price from the government because all the conveyancing water would also be saved. 'It will happen', says Merv. 'It's a monty that 54 kilometres of canal will be decommissioned here within five months. And if we don't get a decent rainfall, another 30 per cent will be decommissioned.'

Already, separate negotiations are underway with a smaller group of about eight irrigators closer to Wakool. The way Merv tells it, the outer spurs of the canal system will be the first to be decommissioned, like a bush that's being pruned back to its trunk and main branches. Unlike selling one year's allocation, these sales of water would be permanent. The canals would be shut down, never to be recommissioned. The blocks of land would be far too small to be viable as dry-land farms; they would be consolidated and the district depopulated. Once started, the process would be irreversible. 'I think a lot of people understand that they're too far down that track. Something has to give … the water is going to leave our countryside. It goes against the grain, it goes real bad.'

And then Merv tells me about Tuesday night. While Andrea and Sam and I were at the show meeting, Merv was at the P&C meeting. Already, the school out at Burraboi has closed and the remaining students have moved into Wakool. Merv reckons it's a great little school, with two teachers for the twenty-five students. One of the them is the principal, a dedicated and professional woman who drives the 120-kilometre round trip from Moulamein each and every day. He tells me what decommissioning the canals would mean for the school. 'Our headmaster spelled it out really simply. We've got twenty-five students here at the moment. If we go to twenty-four, she's got to leave us. Because the school can't support two teachers with under twenty-five students. For the want of one student. Now, if we lose this water, we lose the education for our kids.' Merv reckons that if the twenty-two landholders on the Moulamein spur and the other eight closer to town sell their permanent water and leave the district, then there will only be something like fourteen kids left in the school.

And for a moment the big bloke is quiet. He's got something in his eye and concentrates on wiping it away.

⌒

To anyone but a journalist, the words 'final communiqué' may hold a certain mystique. They suggest foreign capitals, fans revolving slowly overhead, world leaders haggling late into the night over matters of principle. Journalists know better. To them, the words suggest some of the most opaque and convoluted language known to humankind: an unholy mixture of bureaucratese, legalese and spin. So often the purpose is not to communicate what has been achieved, but to disguise what has not, to create the veneer of unity and the impression of progress. I've covered APECs in South America, CHOGMs in Africa and ASEANs in Asia; I've had my fair share of final communiqués. The preambles are worthless, padded with words like 'historic', 'unprecedented', 'forward-looking' and 'united'. The texts themselves are linguistically tortured. But for all of that, they can be valuable reading. The thing to look for is not what is said, but what has been left out.

In July 2008, Kevin Rudd met with the relevant state and territory leaders and signed the Agreement on Murray-Darling Reform. It calls for the development of a basin-wide plan overseen by a new organisation, the Murray-Darling Basin Authority. From 2011, the authority will have the power to set limits on how much water is extracted for irrigation from the basin's rivers. The agreement is riddled with compromise, for it involves states giving up power to the Commonwealth in return for money, a time-honoured piece of constitutional horse trading. But for all the density and ambiguity of the language, it's still possible to detect a strong motivation behind the agreement: the smell of politicians panicking. Just how the needs of irrigation, industry and the environment will be prioritised remains unclear, but not so the rules for critical human needs. Critical human needs means water for drinking, washing and other domestic uses. Read the agreement and it's evident the basin governments are laying the ground rules for an apocalyptic scenario where the basin runs out of drinking water. With three million people wholly or partially dependent on water from the basin, not including Melbourne, this is serious stuff.

'The parties recognise that critical human water needs are the highest priority water use for communities dependent on the water of the Murray-Darling Basin', the document reads. But not all communities. 'The parties agree that the arrangements for meeting critical human water needs for those communities dependent on the River Murray System, excluding the Edward-Wakool System downstream of Stevens Weir … will be a mandatory part of the Basin Plan.'

So every community on the Murray River, from its headwaters in the New South Wales and Victorian Alps, all the way until it reaches the lower lakes at Wentworth, are guaranteed to have priority access to all available water in the Murray, ahead of irrigators, industry and the environment, plus access to water from rivers like the Murrumbidgee and the Goulburn—every community, that is, except for those around Wakool. If the crisis in the rivers worsens, then the residents of Wakool won't even get drinking water, let alone any for irrigation. Somewhere in Canberra, the technocrats have decided that Wakool isn't worth the conveyance water. And if it's not worth it for drinking water, then how can it be worth it for agriculture? I think they want to shut down the irrigation area. But much better to prolong the suffering before stepping in to buy the water. Much better to pick the irrigators off one by one in distressed sales, much better to drive the price down as far as possible. Much better to announce generous sounding transition grants once the farmers are bankrupted, much better than telling them outright that their time is up, much better than buying them out while they retain some of their savings and some of their dignity. Much better for the politicians; much better for the taxpayer.

⁓

It's Friday, show day. Gail Tully has roped me in to judge the photographic competition. 'There's no pressure', she says. 'You're leaving town anyway.' But I don't need to be out at the showground until the afternoon, so I decide to go for a drive. The day has dawned overcast, but the chances of rain are remote. The weather looks like being perfect, with none of the heat of previous years. I drive north-west out of town, along the road sealed by the feedlot company. Twelve kilometres further on I reach the rice silos at Burraboi. There's not a person in sight, not a single car. Just the wind and the metallic echoes of the empty corrugated-iron warehouses. Something is banging somewhere, and a crow lets out a plaintive call on cue. The silos and sheds are massive, so much more imposing when seen from the ground than from the air. The signs are a little faded, but that's all—the machinery, the conveyor belts and the gantries all look in perfect working condition, ready to be reactivated at a moment's notice. I keep driving and ten minutes later I reach the Yambina feedlot, Wakool's lifeline. I smell it before I see it, the unmistakable odour of confined animals. I don't bother trying to visit. There's not much point. I turn the car around and head back towards town, lost in thought.

Something catches my eye up ahead. I start braking, lightly at first, then hard. There's a goose sitting in the middle of the road. A plump white goose. I stop just a few feet short, looking around to see where it has come from. Five other geese look on curiously from the verge. The goose stares at me defiantly. I honk my horn. She honks hers. I honk again, and reluctantly she waddles off to the far side of the road. I'm forced to wait while the other five, one white and four grey, cross the road to follow their leader. I climb out and look around. There's no sign of where they might have come from. I start off again, but only about half a kilometre further along I come to a farm driveway. I drive in, figuring this is where the geese have escaped from, intending to tell the owners.

I come to an old farmhouse surrounded by machinery sheds. From somewhere inside I can hear a radio playing. The farmhouse is overgrown; there's a stillness here, as if nothing has moved for a long time. It takes a moment to decide where the front door is. I push through a half-open gate, past some bushes growing across the path, and bang on the door. Nothing, just the radio yabbering away to itself. I yell a greeting. Still nothing. So I try the door. It's open. I push it wide. It's dark inside, and the smell hits me while my eyes adjust. It's the smell of damp and mould, of a room that's been shut up too long, the disturbing smell of things gone awry. The first thing I see are the piles of clothes scattered about the floor, then the ripped mattress, and then the bottles. Dozens and dozens of empty bottles, some strewn here and there, some in piles, some lined neatly side-by-side on shelves. A tableau of despair. For a heartbeat I'm held frozen, and then I'm backing away, back through the door, repelled by the sense of lives askew, of foreboding and trespass. 'Let's take one more caller …', says the cheerful voice on the radio as I retreat through the gate. I get in the car and drive quickly back out along the drive and onto the road. There is no sign of the geese.

In town I stop in at the store and ask Sandra for a coffee, a flat white. I'm still feeling a bit spooked. When she brings it across to me I recount the story of the geese. I tell her the name of the property. 'Oh', she says. 'There's no-one there anymore. He died.' And she moves into the post office section and starts sorting letters.

Later, in the early afternoon, after doing my duty judging the photographs out at the showground with Gail, I head back to the pub, still distracted by thoughts of the empty farmhouse. Out the back, a young bloke is slouched across a table in the courtyard by my room, too drunk to move, a line of spittle running down his cheek, a lit cigarette dangling from his hand. I decide to have a power nap before the show proper begins, but I'm woken by a knock on my door. It's Chelsea, telling me there are a couple of blokes in the bar I should meet. Inside I encounter 76-year-old Ronnie Newell and his mate Noel Green. The two men have dropped by for their regular Friday afternoon drink. They tell me they arrived in town within months of each other, back in 1950. Wakool was a different place then; the biggest problem wasn't water, it was rabbits. They were everywhere, and every farmer employed a rabbit trapper. A bloke called McGrath ran a series of freezers and would pay two and six a pair, which was good money in those days, and legally tax free. The meat would go off to the city markets and the fur to the Akubra factory. Ronnie says McGrath made so much money, he bought the Queen's Bentley after her 1954 royal tour. The trappers made enough to buy Holdens at £670 apiece. One rabbit drive, with all proceeds going to the Boy Scouts, caught 3000 of the animals. Noel says it was the great floods of 1955 and 1956 that finally got rid of the rabbits, when the new disease of myxomatosis was spread by mosquitoes. He reckons the bloke who invented myxo saved Australia. 'Lot of native trees were killed by rabbits', says Ronnie. 'We can blame the Poms for bringin'' them out. Rabbits, foxes, and politicians. They were the three biggest scourges. But the rabbits aren't so bad nowadays.'

Ronnie's been crook lately, after a life of shearing, farming and amateur boxing. He was good enough with his fists for the coppers down at Barham, Long John Gotten and Jimmy 'Bumper Bar' Dunbar, to ring him up from time to time and enlist his support to break up trouble in Wakool. Ronnie would tell the miscreants to straighten themselves up or he'd deck them. It usually did the trick. The police only had a motorcycle with a sidecar, which made it a tricky business to arrest someone and haul them back along the dirt road to the Barham lockup. So whether it was Ronnie or the coppers, the fist or a good kick up the arse was deployed more often and more swiftly than formal arrest. Despite his run of bad health, you can see Ronnie has been a big man. You can tell by his hands. And there's still the spark of charisma swimming in his watery blue eyes, still the quick retort and cheeky grin that make him a Ronnie instead of a Ron or a Ronald. He reminds me of the cricketer Keith Miller, when Miller was an old man but refused to believe it.

'They went and hanged a bloke out here once', says Ronnie. 'His nickname was the Teal Duck. Back after the Second World War, beer was rationed, and you could only get a couple of bottles … Blokes'd get their couple of bottles and take 'em out and put 'em in their vehicle or whatever. And this bloke was knocking them off. So Ray Peg and another bloke, they caught this bloke and held court out on the footpath, just out there. And then they got a rope and put it around his neck and bloody well hanged him. But they didn't do it properly. His feet were swinging off the ground and he was choking. Jack Gilmour came down from the garage and cut the rope and let him go. They saved him. The last time anyone saw of him he was heading towards Deni, on foot, in a cloud of dust. He was never seen again. That was the Teal Duck.'

Ronnie finishes his beer and I offer him another, but he says he has to go. But before he does, he leans over to me and says, 'When the red gum gets in your blood, it's there to stay. I love the red gum country, the flat country. I hate the bloody hills. Drive to Barham from here and you only change gears once. Live in the city and you do nothing else, using the clutch all the time. No wonder cars last longer here'. I shake his hand and feel the residual strength in the fingers of the old shearer. He's done me a good turn, erasing the despair of the farmhouse, with its empty bottles and fugitive geese.

It's past five o'clock, time for me to head out to the show. I pull up at the footy ground gate where Andrew Tully and Merv Membrey are doing the honours. I pay the $10 entry and get some raffle tickets: three for $5. I park the car; I'm among the first to arrive. I walk over to the clubhouse where I chat with Gail, who is sitting behind a table selling showbags. Inside the pavilion, the exhibits are on display. The photos I judged earlier in the day are mounted on partitions, with the championship ribbon next to a shot of green balloons floating across a lush field of yellow and green canola, a dark line of river gums on the horizon. Annette from the store has picked up the championship prize for a jar of relish. There's a display of Vacola jars filled with preserved apples, apricot jam and Asian plum sauce. The bar in the corner is well stocked with beer, and a couple of workers are getting in early. I run into Mick and ask how it's all going. He's happy. The weather is holding and the bus company is bringing people across from Deni. The ribbons for the horse events have arrived in Barham and Andrea has scooted over to pick them up. I drop by the adjoining canteen and say hello to June. She asks whether I got some lunch when I was out judging the photos. I tell her the cold cuts and salad were just perfect.

Next door to the pavilion, a marquee has been set up containing a petting zoo. I drop my gold coin donation in a bucket and enter. A young girl is having her face painted, but otherwise the tent is still waiting for more people to arrive. A Shetland pony named Snoopy waits patiently in a corner pen, near the results of a colouring-in competition. Kids from the school have brought in their pets and farm animals. A birdcage carries the sign 'Wendy and John are Budgerigars', and another, 'Monroe and Jason are Cockatiels'. 'Blacky and Spotty are Yabbies' says the sign on a bucket of murky water.

The old-fashioned squeal of microphone feedback draws me back outside, where the crowd is starting to pick up and the Miss Showgirl competition is getting underway on the back of Allan Adams' truck. First up are a bewildered bunch of preschoolers competing for Miss and Mr Tiny Tot. Chelsea and Ginger's son, Jake, is crowned Mr Tiny Tot. Grandmother Anne beams with pride. There are seven entries in the Miss Teenage Showgirl: six fresh-faced girls and a boy dressed in a gorilla suit. The gorilla doesn't win but gets a big round of applause nevertheless. The band fires up 100 metres away, protected from the elements by a three-sided tent. It's the three-piece from Deniliquin, youngsters dressed in black-and-white op-shop cool. No-one pays any attention except for a couple of admiring primary school girls. Not far away, a bunch of teenagers is kicking an Aussie Rules ball. One pretty blonde-haired girl launches a bare-foot punt that sails a good 35 metres. The band finishes its sound check and makes a reasonable fist of looking nonchalant while it lounges by the boundary line. They watch as Miss Teenage Showgirl totters by clutching her sash, her high heels sinking into the turf.

I cross the oval to sideshow alley. It still looks a little tawdry in the dimming light of the setting sun, its glowing lights just beginning to radiate the promise of fairground magic. The spray-on tattooist is there, but so far has no takers. Perhaps when people have had a bit more to drink. But the dodgems are already eliciting squeals of delight from their teenage drivers. I see Andrea and Sam squashed into a red car, Sam's mouth and eyes wide with delight as he jerks the wheel first one way and then the other. Andrea hangs on and shouts advice. The octopus ride, 'Australia's Greatest', whirls madly, turning pink faces green, while two inflatable slides, 'Treasure of the Caribbean' and 'Magic Carpet', offer a more sedate thrill. 'Every Child Wins a Prize' says the sign above the revolving clown heads. Nearby, toddler-friendly boats gently circle a small pool of water. A garish stall offers the unholy trinity of dagwood dogs, fairy floss and cream-filled waffles, but it's the cappuccino wagon that brightens my day. Talk about salvation. But I'm its first customer, and the brew comes out tainted with cleaning chemicals. They make me another but the taste lingers.

I return to the pavilion and have a sausage sandwich. A young bloke comes up to me, thrusts out a hand to be shaken and says, 'Dave'. Dave says he's seen me at the pub and offers me a beer. We get talking. He's a good-looking young bloke, with an infectious laugh and an optimistic glint in his eye. I don't know where the optimism comes from. He tells me his parents lost their farm back in the 1990s and split up. He never made it to high school, leaving to look for work when he was thirteen.

'Can you read and write?' I ask.

'I can't write but I've learnt to read a bit during the last few years. It's not such a big problem. You can always find someone to write stuff for you', he replies.

Dave is twenty-four. He's worked at the mines up on Groote Eylandt, and as a laser leveller down near Bendigo. Now he has a job cleaning out pens at the feedlot. It's not great work, but he likes Wakool. He's bought a house for $80 000. 'A brick house', he says. I ask if he has a girlfriend. 'Nearly', he says with a shy smile. 'I'm working on it.' He tells me all the local girls go off to uni, 'even some of the guys'.

Out in the middle of the oval, four local volunteer fire brigades are putting on a demonstration, racing to unwind hoses and put out fires blazing in 44-gallon drums. The sun has set and the fires cast an orange glow over proceedings. The Tullakool brigade springs a leak, spraying water everywhere, much to the delight of the announcer whose voice booms out from the PA: 'Anyone out there from the hippy tribe, now's your chance. Just head over there and get a free shower. Don't be shy!'

'So what are you writing your book on?' Dave asks. 'Why we're all knocking ourselves off?'

'Are you?'

'For sure it happens. Maybe not as much as it used to. I think it was more common back between 1995 and 2005. But you don't really know. They don't put it in the news or anything. Not unless they take someone with them.'

'How do you mean?'

'You know, if they kill their wives and kids as well. There's been about three in the last couple of years. There was one a few months back in Barham. Otherwise, people don't talk about it. You ask "What happened to so-and-so" and they just say "Oh, him. He died". End of conversation.'

The raffle is drawn. I win a prize, then another, receiving a chiacking as I enter the hall to collect my winnings. I select a couple of bottles of wine and a picture frame for my wife, Tomoko, who's juggling work and the kids back in Canberra. The band starts up. The Lincolns. They're not half-bad. I leave my winnings with Dave and head over to the bar to get him a drink. I see June working at the counter of the canteen and give her a wave. I buy the beers, then run into John Licato and Gail. John picks right up where he left off a few days before: the politics of water. Gail listens, brow furrowed. I excuse myself and head back towards Dave. A pretty young girl has joined him. The look in their eyes would melt cement. I try to deliver the drink and make a discreet withdrawal, but Dave is having none of it, enthusiastically introducing us, although I miss her name. She tells me she's at uni in Wagga Wagga, studying to be a schoolteacher.

'How do you like Wagga?' I ask

'I don't. It's too big. Too many people. I prefer it here', she says, and casts a shy glance at Dave.

It's fully dark by now, almost time for the fireworks. I run into Andrew, who has finished his shift on the gate. He reckons there aren't as many cars as for the footy, but they're making more money: 'There's more fifties in the tin'. The band is ploughing on, playing everything from Nirvana to Rick Springfield. Most of the crowd is hanging back by the pavilion, women chatting and men drinking, but a few teenagers are dancing on the grass. Youngsters are running everywhere, high on fairy floss, the girls in phosphorescent necklaces, the boys wielding glowing plastic swords. The band takes a break and the fireworks begin with a whoosh and a bang, each explosion echoed by oohs and ahrrs from the crowd. They're not as high or as mighty as a capital city display, but in some way they're better, bursting low and directly overhead. Intimate fireworks. Last year's warnings have gone unheeded; embers rain down on the band's marquee. Ash and debris float across the crowd, but no-one is complaining, least of all the kids, who stand open-mouthed and enchanted. The display ends with a grandish finale and a round of applause. With the fireworks over, families start drifting away, but The Lincolns fire up again and the bar is doing a roaring trade.

I talk to one of the women who works at the school, a relative of June's. She took the prize-winning photo of the balloons. She tells me what an absolutely terrific school they have and just how hard-working the head mistress is and how disastrous it would be if anything should happen to it. Then I have a long chat with Howie Davison, seventy next birthday, a professional fisherman. I ask how he can be a professional fisherman when there's no water in the river. He holds a yabbie and carp licence, one of only twenty-one in New South Wales. He says carp are worth between 30 and 80 cents per kilo, while yabbies can fetch around $18 a kilo. He reckons the Paroo graziers are right on the money about yabbie gangs operating out of Mildura.

Over by the bar I run into Ben, a 23-year-old shearer who has walked straight off the page of a Henry Lawson story. He went on the road aged fifteen and had a hard time of it. He says he could shear about fifty sheep a day when he started, but he can do 120 now if he pushes it. His average is about eighty. Pay is $2.35 per sheep. His face is a strange mixture, the pimples and fluffy beard of late adolescence blending into the deep lines of experience. There are also a couple of scars. 'Maoris, mate. There was one in a shed who wanted to fight me. You've got no choice. If you don't stand up for yourself, they walk all over you.' Ben's wandered far and wide, moving from shed to shed, but he's come back to Wakool now to look after his mum. He's signed up to play footy again, too. He makes it sound like the biggest commitment of his life.

The crowd is thinning out now, and The Lincolns wind up with an ear-bending finale and a spattering of applause. I do the rounds and say my goodbyes. Mick is satisfied the night has gone well. June is still behind the canteen counter. I ask how she's holding up. 'Not bad, mate. I have to be back here at 7.30 in the morning to make egg-and-bacon rolls for the horse people.' I come across Dave and his girl. He's had a few drinks, but he's also got his arm resting protectively round her shoulders and a big dopey grin on his face. I take their photo and wish them well.

The next morning I get up early and head down the corridor for a shower. I drink a couple of glasses of water, trying to hold off what may or may not be the beginning of a headache, then pack up my stuff. I find Chelsea down in the kitchen and settle my account: $125 for four nights accommodation and two counter meals. I say my goodbyes, walk out the back and climb into my car. I think about going to the footy oval and bidding yet another farewell to the people who have been so kind to me. June will be in the canteen feeding the horse people, Max or Mick or Andrew or Merv will be on the gate, and Andrea and Sam will be riding their horses in competition. But it's a long drive back to Canberra; I've lingered long enough. So I resist the temptation of a flat white at Annette and Sandra's and drive past the park to the T-junction where the main street meets the Barham–Deniliquin Road. I give way to a truck full of sheep who are heading unknowingly towards the feedlot. Then I turn towards Deni and leave town. I pass out of Wakool Shire and into Murray Shire. 'One of the Fastest Growing Shires in New South Wales! Population 7,000' reads the sign.

Drought and circumstance have settled heavily upon this land and its people, so reliant on irrigation but so starved of water. The government is preparing to pick them off one by one, buying their water when they can no longer afford to keep it. It seems a cruel way to go about it. The people of Wakool know what's happening, but that doesn't mean they accept it. They're not going down without a fight. Surely such tenacity must garner some reward, or at least some justice. I drive towards our distant capital and hope it rains. I hope it bloody well buckets down.

8

TO THE RIVERLAND

Headings Cliff, on the South Australian Murray, north of Paringa

My father, Kevin, was born in the Victorian bush, at California Gully near Bendigo. My mother, Glenys, was born on the coast, at Malo, where the Snowy River met the sea in the days before it was dammed and diminished. Men, they say, are from Mars and women from Venus, but I know one thing: my dad is from the bush and my mum is from the beach. As a schoolgirl in Melbourne, Mum would spend her weekends sunbaking down at Brighton Beach. Even now, she loves swimming in the sea, staying in for hours, close to the house she finally convinced Dad they should buy on the south coast of New South Wales, near Moruya. Dad likes it well enough, but when he goes for his afternoon walk, he heads off into the bush, not along the beach like everyone else. It's in their very nature: Mum is intuitive and of the ocean; Dad is practical and of the land. I'm not sure where that leaves me: somewhere in-between I guess. An affinity for rivers is about the best I could have hoped for. I'm a bit like Canberra: an awkward compromise.

My parents met in Murrayville, a portentous name if ever there was one. Mum went there on her first teaching post, assuming it must be on the river. It's not. It's in the middle of nowhere, also known as the Victorian Mallee. The Mallee lies in the state's north-western corner. There's one sparsely populated swathe running east–west below the Murray, west of Mildura and bordered to the south by the 'sunset country' wilderness. Below the sunset country lies another marginal band of farmland, extending from the Murray to the South Australian border. Murrayville lies halfway along this southern band. To the south again lies the Big Desert. The Mallee is home to a tough breed of dry-land farmers, with the emphasis on the dry, who eek out a living in near perpetual drought. Rainfall is about 15 centimetres a year, often less. Only the quality of the soil makes the land viable. Depending on who you ask, the term 'Mallee Farmer' is synonymous either with stubborn resilience or with a pathological ability to bet the farm on the next crop. A hundred thousand dollars, maybe two, can be spent sowing a crop on the rumour of rain. It's not a place for the faint-hearted.

Dad arrived in Murrayville the same year as Mum: 1953, his second year out as a teacher. The town was home to a consolidated school, a then recent innovation that had gathered all the one-teacher schools and used buses to bring in pupils from a wide radius. Dad came equipped with an agricultural science degree. He taught farm science, ran a small farm with his students, and made a few bob on the side fattening a flock of sheep on the school oval. It was a time of postwar confidence; Dad arrived preaching the new religion of technology, DDT and modernity. He advised local farmers on the latest methods and tested their soil for trace elements. He ran agricultural field days, played footy for the local team, and lived out the back of the pub. He made such an impression that when he and Mum decided to leave Murrayville, a wealthy farmer named Heintze tried to hire him to open up farmland on the edge of the Big Desert. With no money of his own, but all the know-how in the world, it must have been a tempting offer. Dad would have loved to have had a crack at farming. But he knew the Mallee could be synonymous with something else: heartbreak. He turned the offer down, much to the relief of my mother.

I drive into Murrayville, 'Gateway to the Victorian Outback', on a warm autumn afternoon. There's been some unseasonable rain, but then any rain in Murrayville is unseasonable. It's settled the dust but has failed to invigorate the town. It feels not so much gripped by drought as by a pervasive sense of fatigue. Stretched out along the highway, the buildings themselves look tired, the long years having eroded that postwar optimism. They sag, as after a long sigh, their paintwork fading in the sun. An old-fashioned garage, bowsers out in the road, stands with its wooden double doors peeling and propped open. 'Renown Garage' reads the sign above the door in lettering once blue but now powdery grey. 'M bil Service' it says, the red 'o' peeled off and blown away. Further along, a low line of shops stands empty below a rickety awning, the rendering along the roofline crumbled, revealing the erratic brickwork beneath. Up on Reed Street, the main street by default, six stores remain open, huddled together under the corrugated-iron awning of a single-storey terrace row, its facade weathered to pastel and brick. There's Thurlow's Newsagency, a Christian bookshop and Nanna's Milk Bar, with a large 'For Sale' sign in the window. The Commonwealth Bank branch, with its wheelchair-access ramp, its aluminium-framed windows and its fluorescent brightness, has been transplanted in its entirety from a fresher, more vibrant Australia. Round the corner, a rusting Austin A30 stands in a vacant lot with a 'make an offer' sign propped behind its windscreen.

In this climate, the rust is the most impressive thing about it. The grill is missing and so are the hubcaps. Perhaps the same car—shiny, black and brand new—once won an admiring look from my father, arriving in Murrayville as part of a postwar generation of Australians surprised to find themselves contemplating car ownership. The A30 was powered by a straight-four engine displacing 800 centimetres, or four-fifths of a litre (perhaps giving rise to a colourful colloquialism). It could rocket from zero to 60 miles per hour in less than a minute. A real tearaway. I wondered if my parents had ever ridden in this car, this very car now gently falling to bits in the vacant lot, perhaps sneaking off to the movies across the border in Pinaroo. I wouldn't be surprised; the whole town gives the impression that not so very much has changed since my parents left. Only the pub, the town's sole two-storey building, is freshly painted and bright. But like the Commonwealth Bank, it's an illusion created by money from 'away', the new veranda and other restorations funded by a government grant. But at least Murrayville is still here. Other towns that existed in my parent's time, like Danyo, 14 kilometres to the west, have disappeared back into the scrub altogether, with only a highway sign to acknowledge they ever existed.

I drive towards the school where my parents taught, but my imaginings are interrupted by a man standing in the middle of the road, waving his arms frantically. It's Ken Heintze. At age eleven, he was one of my father's pupils. He's been expecting me and must think I'm lost, not something easily accomplished in Murrayville. I abandon my meanderings and pull into the driveway of his neat brick home. Ken is the nephew of the man who once offered to stake my father. He introduces me to his wife, Raelene, and then asks if I want to see the farm. Ken, now sixty-seven and retired, still helps out on the family property most days, driving a tractor for his son or taking a semitrailer through to South Australia. He's a thin, fit looking man, his grey hair and grey beard resting easy on a lean face under the sort of baseball-style cap favoured by American farmers. Only later does he tell me he's diabetic and has to keep a close eye on his blood sugar. It's hard to believe; he displays none of the lassitude of his town.

Maybe it's the rain that has put a spring in his step; it's the first decent fall in five months. But the lack of rain before now hasn't bothered Ken at all. Just the opposite. He says it's too hot and dry to grow a summer crop in the Mallee, so rain in the hot weather does nothing but germinate weeds, and weeds cost money to control. I smile at the paradox: the worst drought in a century, and here's a Mallee farmer happy that it hasn't rained for almost half a year. Now however, rain has come, right on time. It puts a smile on Ken's face as he casually drives at breakneck speed along dirt roads, one hand flopped atop the steering wheel. We're at the farm in minutes. Ken and Raelene's son, Mark, is on the tractor, sowing wheat. Or rather he's *in* the tractor. It's a huge thing. He can't stop to talk; instead his voice crackles across the u.h.f. radio. I can see him inside the cab, peering out the back window, watching the sowing machine. It's not a plough, nothing so crude. Rather, the seed is being drilled directly into the soil using compressed air, causing little dust or loss of moisture. Mark's hands are nowhere near the steering wheel; the tractor is computer controlled, guided by satellites floating in low orbits. The town may have barely changed, but out on the land the technological revolution has continued unabated, rendering medieval the science of my father's day. For the farmers it's a constant struggle: forever adopting new technology, forever expanding to achieve economies of scale, forever investing to stay profitable. It's a struggle Ken fears Mallee farmers are losing.

'It's a build up over a long time, since the late 1970s', he says. 'Our expenses have started to get level with our income and in many cases gone beyond. You can only last so long by running at a loss. But we know nothing else and this is all we want to do, so we try and do it better. It's not easy at all.' Some years ago a prominent politician, a farmer himself, told Ken that entire towns the size of Murrayville were being wiped from the map in the United States. 'I said we're not going to let that happen. Or at least we're not going to let that happen one minute earlier than it has to be. There's lots of unknowns in life, and not many guarantees. You try and do things to the best of your ability. There's a lot of stresses on rural people at the moment. And lots of debt. But it's still a good life. And if I ever had to leave this place it would be very difficult, because of all the people I know and like.' After a lifetime of aridity, it's not drought or climate change that worries Ken; it's the decreasing margin between costs and returns.

We jump out of the truck to clear some old barbed-wire fencing that has somehow been dragged onto the paddock. We're on a slight rise and it gives me a chance to take in the landscape. The Mallee isn't flat, not the true flatness of Wakool and the Riverina. It's not exactly hilly, but there is a rise and fall. It's a land of whispered promise, a land that needs nothing except rain. It must be something to behold, that one year every decade or two when the heavens really open, and the monotonous tans and shallow creams are illuminated with green. A river wouldn't go astray either, but they've never had one here so I guess they don't know what they're missing. We take the barbed wire over to a farm dump, near the scant ruins of a farmhouse, a relic from when the land was settled. The Mallee was largely populated by soldier settlers after World War I. Perhaps after the mud of Flanders, the Mallee looked appealing. The diggers were given 680 acres of land— 1 square mile. And dig they did, undertaking the brutal work of hand clearing the land of the deep-rooted Mallee scrub. But 1 square mile was never going to be enough. They fought on in isolated poverty, surviving on mutton and flour, as the unforgiving land blasted their bodies, tormented their minds, and sniped at their souls. The only way to survive was to buy more land, so neighbour preyed on neighbour in a grim Darwinian dance. By the outbreak of the next war, barely twenty years later, some 60 per cent of soldier settlers in Victoria had walked off their land; the attrition rate in the Mallee must have been among the highest. But it meant that those who survived were the best of them: the shrewdest, the hardest working, the luckiest. Ken Heintze's grandfather settled near Murrayville in 1924 and prospered, and so did his sons and their sons.

Now a new threat has emerged. Or rather, a new temptation. Murrayville's hidden strength has always been that it sits above an aquifer of pure water, water that stretches off westward across the South Australian border. It's why the town was founded far from a river, and why it has endured. Even in the worst drought, there's been water for stock and for people, water for the town. Now the irrigators have come, refugees from parsimonious rivers, frugal canals and shrinking allocations, alert to the main chance, bringing their massive centre-pivot irrigators with them—huge spider webs of steel 100 metres or more across that revolve on motorised wheels round a central pump site. Ken Heintze doesn't like them.

'The beginning of irrigation in the Mallee was about ten years ago', he says. 'It's grown to a point now where it is very prevalent. They're growing potatoes, mainly. There's lots of centre-point irrigators pumping huge amounts of water. And as a dry-land farmer I wonder whether it's sustainable. What if it's not? Or if it reduces the quality of the water? That could have disastrous effects. We rely on underground water … If the quality of the water deteriorated so it was undrinkable or couldn't grow anything, like vegetables, that would be drastic. The irrigators could possibly move on, but for those of us who live here, it could spell a catastrophe.'

Ken's comments cause me to once again revisit the concept of a river. I've long abandoned the *Oxford* definition: 'Copious stream of water flowing in channel to sea …' The river has expanded in my mind, encompassing first the flood plain and later the entire catchment, so that it has become the culmination of water in the landscape, not something separate from it. But if my understanding of the river has stretched from one dimension to two, from a line through the landscape to the landscape itself, then why not to three? Water flows not just across the top of the land, but through it as well. A full river channel will push water out into the subsoil; its pressure will resist the inflow of underground aquifers. It's a matter of balance: if the level in the river channel is low, an aquifer may spill into the river, carrying salt with it. But if the aquifer itself is depleted, then the river may leach away into the soil. And if water is taken from deep underground, does this mean the water in higher aquifers will flow deeper, depriving some other place hundreds of kilometres away of its water? The new basin-wide plan, to be unveiled in 2011, will not just put a new cap on diversions of surface water, but will for the first time include limits on ground water. It's a move in the right direction, but I can't say I'm filled with optimism. If governments haven't been able to properly manage water above ground, if they've overallocated what they can see and touch and measure, how are they going to manage the mysterious underground flows? You may not know if you're tapping last summer's rain or a downpour from a million years ago. Deplete the aquifers and the blotting-paper catchments may become even more porous.

Later, Ken and Raelene and I go down to the pub for dinner. The first fire of the year is burning in the grate of the front bar, and there is nothing faded in the old hotel's character. The walls have been painted a deep red and the architraves a jolting green, yet the floorboards are stained a satisfying brown and the bar is well weathered and original. Various denominations of money hang from the 15-foot ceiling, ingeniously launched from the floor below, evidence that revelry remains a possibility. But there's little of the self-conscious clutter of tourist hotels; instead there's a kind of appealing minimalism. Photos of the local footy team adorn the walls, together with pictures of a local girl who grew up to be an Olympic basketballer. My father lived at the back of this pub during his time at Murrayville, among a colourful troop that included a couple of alcoholic road workers, a Gipsy Moth pilot known as 'the Black Prince', and a hunchback. I wish he were with me now. I'd like to have a beer with him and listen to stories of Murrayville as it was, back before global warming, the global financial crisis, and globalisation. Back when everything seemed more possible.

Mum, of course, could not live at the pub. She couldn't even enter its front bar, and was too respectable to be seen in the ladies lounge. Drinking in the 1950s was a neatly compartmentalised affair. Men and women were separated, races divided. And there was another divide as well. In Murrayville, as in many towns, there was a wine saloon next to the pub. The saloon sold cheap port and vicious spirits to the drifters and ex-servicemen who floated through the bush, mocking the postwar prosperity. Dad never once entered the wine saloon. He was no snob, but he was a schoolteacher, a respectable member of the community. For him to drink in the saloon would have been as scandalous as Mum sinking pots in the front bar.

Mum lived in a hostel for woman teachers. One of her old colleagues, Ena Lackman, joins us for dinner at the pub. She was the home economics teacher when Mum and Dad were there. Home economics; a title from another era. Ena tells me she came to Murrayville intending to stay for six months, but married a farmer instead and never left, even after he died. I look at her across the table, nattily dressed and smiling, a pleasant face under tight grey hair. It could easily have been my mother sitting there, if things had been different.

Ken tells me he still thinks there's a future in the Mallee, that every year for fifty years, drought or no drought, his family have been able to coax a crop from the reluctant soil. It's often not much of a crop, but a crop all the same. He says not many regions, including the much-vaunted Wimmera, could boast such a thing. No doubt he's gilding the lily, but he has a point. The irrigators have always had it laid on, cheap water on demand. By contrast, the Mallee farmers have been case-hardened, tempered through the decades by the hot Australian sun. They've adapted to the unforgiving climate, the vagaries of global markets, the relentless pressure to modernise. The survival of the fittest. But climate change projections place the Mallee right at the epicentre of the great dry predicted for south-eastern Australia. From little rain to practically nothing. Some things you can't adapt to. I consider its prospects, and Murrayville grows a little more tired and fades a little further into the past.

It's mid-April and I'm on my last venture along the rivers. I've fired up the old Hyundai and it's carried me out from Canberra once more, taking me towards South Australia and the end of the Murray, to where the river is most diminished and its health most parlous. A journey to the end of a metaphor. The road has greeted me like an old friend, taking me out through Yass to the Hume Highway. The sun has been warm, but slants lower across the sky, setting the countryside in glowing relief. The vines at Murrumbateman, just outside of Canberra, were turning yellow and russet as I drove through, the harvest completed for another year, smoke drifting from a fire in the corner of a field. The countryside had looked tired and dry after the long summer, but not noticeably worse than any other year. There's been significant rain up in northern New South Wales and Queensland.

Bourke copped 20 centimetres in one day in February, flooding the town with water and irony. Further to the north and west, an immense tide is washing through the channel country towards Lake Eyre. There's talk the drought may be breaking in the north. Which only serves to accentuate the pain in the south.

Little water has reached Menindee, the stopcock between north and south. The lakes have risen from 7 to 15 per cent capacity. And that's the best of it; across the Southern Basin, the drought has deepened. Rain has fallen here and there, but the rivers and their feeder dams aren't getting any of it. Inflows into Dartmouth and Hume have fallen to their lowest levels on record for the first three months of the year, lower even than the disastrous lows of two years ago. The remaining water in Dartmouth is so jealously guarded that not enough is being released to flush the river: blue-green algal blooms are being reported along the several hundred kilometres between Albury and Swan Hill. The new Murray-Darling Basin Authority has declared that the drought's persistence and severity is unprecedented. The ACT Government has unveiled plans to extract 20 billion litres of water annually from the Murrumbidgee to bolster Canberra's water supply; the announcement has elicited none of the protest ignited by a similar scheme to siphon the Goulburn River to supplement Melbourne's reservoirs.

Autumn is halfway through, but the sky is cloudless and the air warm. I'm wearing shorts and a T-shirt. The Hyundai's driver-side window is open, the rush of the wind laying a cover of white noise across the music from the stereo, as I travel the Hume and climb through Conroy's Gap. From the elevation of the gap, the horizon opens up and I can see off towards the wide flat plain. The distant hills are misted in an orange haze. It could be dust blowing in from the central deserts. Or it could be topsoil lifted by the ploughs of a thousand fields, as farmers set themselves for winter crops and another throw of the dice.

I pass the Dog on the Tuckerbox without stopping. Once is enough. But at Gundagai I take time to check the condition of the Murrumbidgee, the second most important river in the Murray-Darling Basin. It's flowing bank-to-bank, bottle green and vigorous, even now the irrigation season is abating. The Darling may loom larger in the consciousness, and Bourke may capture the imagination in a way Wagga Wagga never will, but in every other way, especially economically, the Murrumbidgee trumps the Darling. It may even be a few kilometres longer than the desert river, depending on how you measure the Darling. But it's the water that makes the real difference: in an average year the Murrumbidgee carries 4250 gigalitres; the Darling carries 12. Only the Murray carries more. Which makes me wonder why so much noise is made about Queensland irrigators sucking up water that might otherwise flow to South Australia, but we hear so little about the major tributaries of the Murray.

Like the Murray, the Murrumbidgee rises in the Australian Alps. Like the Murray, it's fed in part by the Snowy Mountains Scheme. In a non-drought year, the scheme discharges almost as much water into the Murrumbidgee as it does into the Murray. Indeed, the Snowy Scheme is really two loosely interconnected systems: the southern scheme feeds the Murray; the northern scheme flows into the Tumut and from there into the Murrumbidgee. In this drought year, the Murray gets a much larger share of the scheme's water, but then again, the Murray is burdened by downstream obligations that the Murrumbidgee is not obliged to carry. I've looked up the figures: the Murray River storages at Dartmouth and Hume are 14 per cent full; the Murrumbidgee storages at Burrinjuck and Blowering are 34 per cent full.

An hour later, I arrive at Wagga Wagga, its shopping precinct challenging Albury's in size. The river is still full, still flowing, little different than at Gundagai. Another hour or so later, at Narrandera, it's down a bit, but still much the same. I take some photos, and an old man alerts me to a koala curled asleep in the nook of a river red gum. But by the time I reach Hay ('Real People, Real Experiences'), another two hours west, the river is well down in its banks and moving sluggishly. Its green has given way to khaki. It's not unexpected. Both the Murrumbidgee Irrigation Area to the north and the Coleambally Irrigation Area to the south feed off the river between Narrandera and Hay, with huge quantities of water diverted down canals by the Berembed and Gogeldrie Weirs. By the time I cross the vast emptiness of the Hay Plain and arrive in Balranald, the river has turned brown and sleepy, its vitality drained by irrigation, its impetus depleted by the flatness.

I wonder what happens to it next and go in search of the river's end, down where it joins the Murray near Boundary Bend, between Swan Hill and Mildura. It's the junction of Australia's two mightiest rivers, but there's nothing to mark the confluence: no Khartoum, no Phnom Penh, no St Louis. Not even a Wakool. Wentworth sits where the Darling meets the Murray, and Echuca where the Campaspe joins, but there is no such recognition for the Murrumbidgee. Paddle steamers once travelled up the river as far as Gundagai. Plans were drawn up to insert nine locks and weirs on the lower river, extending the inland sea many hundreds of kilometres closer to the mountains. But no port was established at the junction, no settlement, no bridge. There's just a rough and little-used bush track leading through a state forest—a fisherman's track. No other vehicles have passed this way recently; there are no tyre prints in the soft grey silt of the flood plain. There's a puddle or two leftover from some passing cloud burst, and I take it slowly, not wanting to repeat my Barmah bogging.

The junction is beautiful. The red gum–shaded clearing is deserted and serene. I cut the engine and the bushland peace folds itself around me, silent except for the sighing breeze and the piping chatter of bird call. The rivers lie low, well below the heights of their banks. I stand on the Murray's southern bank, looking across to where the Murrumbidgee, about 15 metres wide, enters it at a right angle from the north. A quivering eddy in the Murray betrays the slightest of flows, but the Murrumbidgee is still, its surface like glass. There is no difference in the colour of the two streams, nothing to indicate that the Murrumbidgee is flowing into the Murray at all. If anything, the Murray appears to have backed up into its tributary. It was the same back in early March, when I went searching for the Goulburn River's junction with the Murray in the bushland east of Echuca. The Goulburn, the third most plentiful river in the basin, had slowed to practically nothing by the time it reached the confluence.

I sit down on a pile of dead gum leaves among the stillness and ponder the deserted meeting of the Murray and Murrumbidgee.

Sitting there, alone with my thoughts, I begin to revise my metaphor. I've been thinking of the river as a uniting strand, the thread connecting the different landscapes of our lives: the mountains, the forests, the deserts, the lowlands and the sea. But it's an old metaphor, based on an old river, the river of the imagination. That river is no more, its crystalline waters polluted by politics, commerce and ideology. In the real world, it's been dammed, managed and diverted. It's been measured, divided and allocated. It's been bought, sold and argued over. In the real world, the modern world, as in the ancient, rivers are borders, dividing more often than they unite. I've witnessed a few of the nastier ones: the Jordan River, encased in military no-go zones; the Spree in old Berlin, the wall snaking along its bank; the Neretva in Bosnia, dividing Muslims from Catholics in Mostar. I think of the Styx, dividing the living from the dead, pennies placed on the eyelids to pay the ferryman. But why is there next to no water flowing from the Murrumbidgee into the Murray? Is this also about borders?

I approach the George Chaffey Bridge, heading across the Murray from New South Wales to the Victorian garrison town of Mildura. 'The most liveable, people friendly city in Australia' says the vision statement. Off to the right I spy an oversized replica of Rameses II. The pharaoh sits among palm trees, championing a motel. I add him to the list: the Cunnamulla Fella, the Dog on the Tuckerbox, the giant Murray Cod at Swan Hill. The god-king makes as much sense as any of the rest. Maybe more. The land of the pharaohs was utterly dependent on the Nile, and the thin green strip of enterprise stretching downstream from Mildura is not so very different. Maybe Mildura could strike a sister–city relationship with Aswan. I'm sure the councillors would enjoy the junket; they always do.

At Mildura, home to lock eleven, the inland sea proper begins, kept at a near constant level, stretching almost 900 kilometres into South Australia through another ten locks and on to the river mouth near Goolwa. Hire a boat and you can putter down to the ocean and back, no matter what the season, no matter how severe the drought. The water level at lock eleven is just 35 metres above sea level, but it never changes.

It's not just Rameses II that puts one in mind of Egypt. Mildura casts itself in the same mould, promoting itself as 'an irrigated oasis in the midst of an arid land'. Its streets are lined with palms, its air is desert dry. The story of Mildura is the story of Australian irrigation. It's the place where the use of the river shifted from drought-proofing farmlands to opening up new regions, to 'making the desert bloom'. Its story starts with Alfred Deakin, later Australia's second prime minister, who travelled to California in 1884 to investigate irrigation at the behest of the Victorian Government. There he met a couple of Canadian entrepreneurs, William and George Chaffey. The brothers were growing rich by buying land and water cheaply, developing irrigation schemes, and then selling subdivisions. The Chaffeys, bewitched by Australia's potential, sold their Californian holdings and headed for the Murray. The brothers purchased a moribund pastoral station and set up the Mildura Irrigation Colony. Stroll up Mildura's palm-lined main street, Deakin Avenue, designed by the Chaffeys, past the statue of William Chaffey to the Alfred Deakin Centre, and the helpful staff will tell you all about it. Well, possibly not the bit about the irrigation colony going bust. And probably not the bit about the subsequent royal commission holding the Chaffeys responsible. And certainly not the bit about the Melbourne press branding them a couple of Yankee real estate hustlers. But they'll tell you the good bits.

The Chaffeys were an unorthodox mixture of hands-on practicality and utopian vision, of free-enterprise zeal and community spirit. They weren't just interested in growing food and making money; they wanted to build something enduring. They drew up a town plan, resplendent with broad thoroughfares, parks and churches. They funded an agricultural college. It was to be a brave new world: a profitable world, but a better one. And to help it become so, they established Mildura as a temperance colony. On the high ground overlooking the river, what is now called the Grand Hotel was built as the Mildura Coffee Palace, sited right next door to the Chaffeys' own headquarters. But the architects of the Coffee Palace, who perhaps knew a thing or two about Australians, had the foresight to equip it with deep and capacious cellars, no doubt to keep all that coffee fresh. In 1915, the proprietor, a Miss May Williams, obtained a wine licence, which became a full licence four years later. The cellars were put to good use and the local temperance league was left sipping lemon tea. The hotel went from strength to strength and eventually engulfed an entire city block, including the Chaffeys' headquarters.

By this time, the brothers had more to worry about than the demon booze. For reasons best known to themselves, they'd selected an area for development 260 kilometres downstream from the nearest railhead at Swan Hill. When the river ran dry, as it inevitably did, they faced insurmountable problems getting Mildura's fresh produce to market. Dissent blossomed among investors. Three thousand of them had come to live and work in the colony, many arriving directly from Britain. In 1896 the Chaffeys were bankrupted. A royal commission blamed the two brothers for the colony's failure. It was all too much for George, who returned to the United States where, among other things, he helped divert the Colorado River for desert irrigation. However, William, or WB as he was widely known, stayed on and saw Mildura mature into a commercial and civic success. He helped develop the dried fruit industry as a mechanism to overcome the distance to markets—to this day, older Australians associate dried fruit with Mildura's 'Sunraysia' brand. The necessity to preserve the district's produce also led WB to establish first a winery and later a distillery, demonstrating a certain nimbleness of mind for the founder of a temperance colony. He collapsed and died at his distillery on 4 June 1926. I wonder about his dying thoughts, whether he believed the location of his demise apposite, and whether he supposed he was heading towards a well-watered paradise or somewhere altogether hotter and drier. Perhaps it would be fair to conclude that his adopted country had changed him, just as he had done so much to change it. Certainly he was honoured in his own lifetime, elected Mildura's first mayor when it gained urban status in 1920. By the time of his death, he was widely admired for his guts in 'seeing it through'.

The legacy of the Chaffeys can still be found in and around Mildura. The jetsam of the original empire is scattered around the modern town, well preserved by the desert air and heritage-minded citizens. WB's grand house, Rio Vista, now a museum, sits on the cliffs above lock eleven. The name is almost accurate: from the house you can see the tops of the river red gums, but not the water. The winery, the distillery and the Grand Hotel are all still here. The hotel remains the town's social hub and tourist centre, with an epicurean restaurant housed in those well-proportioned cellars, and a pizza-and-beer pro-duction line spilling into the street above.

At Psyche Bend, a few kilometres upriver from Mildura proper, the original pump house has been restored to working order. The Chaffeys' scheme relied on steam pumps, not a diversion weir, to get the water from the Murray. The irrigation channels themselves were gravity-fed, but first, water had to be hoisted an unprecedented height from the river into Kings Billabong. George designed the pumps himself, massive things, based on marine engines. The manufacturers in Birmingham were sceptical of the blueprints and initially declined to have their brand on the pumps. Only after they established that the pumps worked was the brand name attached with retrospective pride. The huge machines have been restored, inlet pipes still dipping into the Murray. A band of enthusiasts, retired gents in overalls and Casey Jones caps, fire up the neighbouring boiler a couple of times a year and run the pumps, lifting water once more into the billabong. The pumps are beautiful in their way: the steel, brass and glass-faced gauges of the industrial age. But it's the elegance of the pump house itself that impresses. It has the dimensions of a church, with a lofty ceiling rising many metres above the pumps. The volunteer taking donations at the door attributes the steep pitch of the roof to the snowdrifts of the Chaffeys' native Canada, and the height of the building to the gantry crane that runs along the top of the walls. Practical reasons proffered by a practical man. But I look at the light filtering down through the delicately arched windows set high in the brick walls, windows that in another setting would house stained-glass, and I can't help but think that George wanted his building to be beautiful as well as practical, a temple for his new and better world.

From the pump house I drive back into New South Wales over the George Chaffey Bridge, giving Rameses II a cheery wave as I pass. I turn onto the Silver City Highway towards Broken Hill, en route to a salt reclamation scheme. Or at least that's the intention. I'm diverted by Orange World. I can't resist the name, not when it's combined with a giant smiling orange set atop a pole. At some time in this country's recent history, someone imbued with entrepreneurial zeal befitting a Chaffey has driven round Australia offloading these giant steel spheres. Painted as apples, oranges and other approximations of fruit and vegetables, they've been hoisted aloft at farms all over the country, advertising the produce for sale below. Paint them white, whack in a few dents with a pin hammer, and they serve just as well outside a golf course.

Inside Orange World I'm greeted by a panoply of orange-related products. There are oranges, of course. And orange candy and orange soap and orange candles and orange gelato and patented orange peelers and orange blossom honey and orange marmalade and orange grove tea towels and every other orange thing you can think of. It's a Rajneeshi's wet dream. The shop's packing-shed origins and home-spun decoration are immediately endearing. And so when the owner, Mario, emerges to spruik a guided tour of Orange World, how can I possibly resist? Mario has realised that hauling tourists around his citrus grove is more fun—and more profitable—than growing oranges.

He's a natural, which is fortunate, because an orange grove is not inherently a scenic extravaganza. 'Cor, this bloke could talk under wet cement', a fellow tourist whispers in admiration as we climb into the carriages that Mario has hooked up behind his tractor. There's myself, a couple of families and a honeymooning couple. Who said romance is dead? Every carriage has a couple of loudspeakers and Mario keeps up a running commentary from the tractor as we wind our way across the red soil of the orange grove. He tells us he has 10 000 trees, that Valencias aren't worth growing, and that imports from California and Brazil are killing the local industry. 'Make sure you buy good Aussie fruit', he implores. Water is not a problem here. It's good to find somewhere where it isn't. The property is on high-security water and has received a full entitlement this year. It's pumped straight from the river. Mario says rainfall is also important, even on the very edge of the desert. The trees respond to rain, gaining something not carried by the river water. But if water is not a problem, not this year anyway, there are other burdens to bear. Costs are high and income low.

'If you are thinking of buying a citrus farm, I would say to you "don't do it" because there is no money in oranges. But if you do, make sure you have at least 10 000 trees and plenty of water', advises Mario. He shows us where older, less productive trees have been bulldozed and new, more profitable varieties planted. Mario himself pitches in on occasion to help pick the fruit. He reckons if he were getting paid, he'd earn about $25 an hour. He looks sternly at the children on the tour and tells them to become lawyers instead, winning approving looks from their parents.

Mario's best story is about the cordial commercial filmed on his property. The first problem was the colour of his fruit: it wasn't orange enough. So the advertising people brought in plastic oranges by the caseload and taped them to the trees. Then they decided the trees themselves were inadequate, so they took down all the plastic oranges and taped them to the avocado trees instead. Apart from that, Mario reckons it was a cracking good commercial. But soon I stop listening to what he's saying and start listening instead to how he is saying it: his accent and his idioms. Despite his name, Mario came to Australia from Belgium, able to speak only French. But any French influence has all but gone. He speaks of 'vino', not 'vin'. He has an accent ripe with the overemphasis of the Greeks, the musicality of the Italians and the hard nasality of the Eastern Europeans. It's a Murray Valley ethnic patois. And instead of omitting syllables or half-swallowing them, as the French do, he overemphasises each sound, and even inserts a few extra. He speaks of the 'the migh-ee-tee mah-ree reev-ah'. The words 'Murray' and 'River' drop a full third of an octave from their first to second syllables. I could score the phrase on sheet music.

The Australian Nile, the inland sea. Began by utopian North American real-estate entrepreneurs, their vision inherited by a population of immigrant dreamers. Perhaps not exactly the new world the Chaffeys had in mind, but a new world nevertheless.

⌐

From Mildura I head west towards South Australia. Once I leave the river's green strip, the land returns immediately to browns and blonds, sun-bleached and bone dry. This is the northern Mallee. The blue sky is patterned with the wispy feathering of high-altitude clouds, lacking the substance to cast shade. The land has the same gentle rise and fall as around Murrayville. I top a crest in time to see a giant willy-willy raising dust from a roadside field. But I'm mistaken: the dust cloud lacks the necessary swirling symmetry. I spy a hard edge amid the russet cloud. Drawing alongside, it lifts for a moment, revealing a ploughing tractor. No direct-drilling here, just old-fashioned tillage, sending the topsoil off into the atmosphere. This was what I could see from Conroy's Gap: a rumour of rain, another Mallee farmer going the punt.

At the South Australian border, a checkpoint in the middle of nowhere, a fruit fly inspector goes through the back of the car with all the civility of Russian passport control. She communicates entirely through grunts. 'Grunt grunt' (open the door). 'Grunt grunt' (and the boot). I wonder what would happen if I were caught with a contraband apricot. Extraordinary rendition? But everything is in order and the 'all clear' is grunted. The strangest thing about the border, however, is the giant black tyre arching over the highway, as if spanning the starting grid of a grand prix circuit. Plastered across it are large letters: 'DUNLOP welcomes you to the phylloxera free Riverland and South Australia'. The connection between the absence of grape disease and motor racing is not explained. The arrow-straight Sturt Highway is temptation enough to speed, and I wonder if South Australia, with its notoriously narrow economic base, has happened upon a new revenue-raising scam: encourage motorists to speed by festooning the highway with motor-racing paraphernalia, and then fine the bejeezus out of them. But as I travel further west, instead of speed cameras, the roadside displays sign after sign warning drivers of fatigue. The signs are in plague proportions, the dominant feature in an other-wise featureless landscape: 'Power naps save lives'; 'Open your eyes—fatigue kills'; 'drowsy drivers die'; 'A microsleep can kill in seconds';

'Yawning? Take a nap'. I count four along one 2-kilometre straight. Here, then, is a state with the humility to concede it's dead boring, that there is little to keep the long-distance driver awake. You wouldn't get that in the 'premier state'. In actual fact, I like the landscape. I like its scope, its consistency, and the bone-white colour of its fields, devoid even of the residual gold of the eastern states. It's been a while since I've seen it, yet it speaks immediately of South Australia. I wonder what it is, this landscape leached of colour: some trick of the light, some characteristic of the soil?

I arrive at Paringa, 'Bridge to the Riverland', and regain the river, which has looped its own way across the border further north. The vision statement is there but not the elevation. I can understand why. Like the engine of an Austin A30, there's not a lot to talk about. The Murray at Paringa, a stone's throw from the Victorian border, with another 560 kilometres to travel to the ocean, is just 16 metres above sea level. The town is home to lock five and is the start of the Riverland, the long, continuous swathe of vineyards that runs from here to the lower lakes. They stretch from horizon to horizon, row upon row of chardonnay and chablis, malbec and merlot, cab sav and sav blanc. Maybe that's what's happened to the river: like misguided latter-day messiahs, we've turned the water into wine.

I leave the highway at Paringa and head north to visit the river before it's totally enwrapped by agriculture. Twenty minutes later, I'm surprised by what I find: the river has changed. No longer does it meander across the flatness with nothing but the nuances of gravity to guide it. Instead, it flows along the bottom of a high cliff of stratified red and yellow sandstone. It has cut its way deep into the rock, the cliff forming a boundary many times the height of the Cadell Tilt. The precipice runs only on my side of the river. On the opposite shore, the land slopes more gently, with the river moving off into the anabranches and billabongs of the Chowilla flood plain. From the Headings Cliff Lookout, I look down upon the trees along the opposing shoreline. Adjacent to the river, they look rich and healthy, but as I raise my eyes towards the horizon, the landscape turns a disturbing grey. Nevertheless, the opaque olive green of the river below looks hale enough, even if there is no perceptible flow. A couple of houseboats, each the size of a suburban home, bask against the shore.

Heading back towards the highway, I brake sharply to avoid a brown spot wandering across the road. The spot reveals itself to be an echidna. It heads towards the edge of the bitumen with the same self-important waddle employed by its cousin, the wombat. I scramble for my camera, hoping to get the kids a snap before the anteater buries itself in the red soil. I'm in luck; I jump out of the car to find it's wandered back onto the bitumen and can't dig itself in. Instead, it shuffles away from me, its quills erect and blond-tipped, as if returning from the hairdresser. I reel off photo after photo, and then guiltily stand guard against passing traffic until it regains the shoulder.

⌒

Renmark, 'Home of the National Rose Collection', challenges Mildura's claim as the Murray's oldest irrigation system. Back in 1886, South Australian Premier Sir John Downer got wind of the Chaffeys' plans for Mildura and scurried across to Melbourne to gazump the Victorians. His timing was excellent. A couple of wet blankets in the Victorian Parliament thought that selling huge tracts of crown land to foreigners at knock-down prices might not be entirely in the public interest. The sticklers insisted the sale be put to tender. William and George, never backward in coming forward, accepted Downer's offer of 100 000 hectares of riverside land. Then, when no competing tenders for Mildura were forthcoming, they grabbed that as well. A third brother, Charles, was imported from California to oversee the South Australian venture, based at Renmark. The two irrigation colonies grew in tandem, although Mildura was always the larger of the two.

I stay at the Renmark Hotel, an imposing, three-storey, art-deco edifice with sweeping views along the river. Not that I have much choice: it's the only hotel in town. The original hotel was constructed in 1891, but like the Mildura Coffee Palace, it was not permitted to sell alcohol. Renmark, too, was founded as a temperance colony. The inevitable advent of numerous sly-grog outlets led the town council to implement a novel harm-minimisation strategy: when the Chaffeys went belly up, the council bought the hotel and granted itself a liquor licence. Demonstrating more business acumen than the Canadians, they then neglected to issue any more liquor licences, establishing a monopoly that remains largely intact today. There's only a licensed club or two to offer competition. The Renmark Hotel claims to be the first community-owned hotel in the entire British Empire. The town itself claims to be the first settlement on the Murray. Neighbouring Lyrup claims to be Australia's only village settlement. Such sweeping, unqualified claims; I love them. I once drove a road in irony-free California where I was confronted by a huge sign declaring 'World's Most Beautiful Freeway'. Perhaps my liking of such declarations comes from spending too much time around politicians, who are adept at qualifying the meaning out of every utterance. Regardless as to whether it was the first in the British Empire, I'm impressed by the hotel's expression of unconstrained bolshevism.

I decide it's my scholarly duty to eat at the hotel's bistro, its terrace open to the street and commanding views of the river. As I sit out in the warm of the evening, the river has the same Pavlovian effect it had in Yarrawonga: I order fish and chips, together with some local wine. While I await my meal, I wonder if Renmark's model of latter-day socialism might not prove a winner elsewhere in Australia. After all, if profits are to be made from alcohol, not to mention the pocket-stripping depredations of poker machines, then perhaps it makes sense to have the money flow to the same authorities that are required to clean up the resulting human flotsam. Then I remember one evening in Russia. A friend had explained over several rather strong drinks one of the reasons why Mikhail Gorbachev was so unpopular among his countrymen. It seems at the height of perestroika, Gorby decided to wind back the state-controlled vodka supply. He wanted to combat alcoholism, absenteeism, workplace injuries and low productivity. His tax office decried the diminution of its one reliable source of revenue, while the proletariat decried the diminution of its one reliable source of oblivion. Gorby never quite recovered after that, especially once Boris Yeltsin, a man who knew his way round the inside of a Stolly bottle, climbed on top of a tank. I'm still pondering these competing arguments when the fish arrives. The two pieces are identical, fish-like shapes, as if expelled from a jelly mould. They're battered in something resembling Soviet-era cement, they taste like relics from the cod wars, and their texture is of reconstituted clag. The wine, made on a local family-owned vineyard is, by contrast, excellent. Free-market capitalism 1; central planning 0.

And yet there is something deeply socialistic about the development of Australian irrigation. After the failure of the Chaffeys' private scheme, the South Australian Government handed control of it to the Renmark Irrigation Trust. The trust's headquarters still reside in the same elegant stone building, veranda condescending, where it has been housed since 1893. The trust owns and controls the water, and the landholders own and control the trust. Water rights are tied to the land—sell the land and the water goes with it, along with membership of the trust. Rights to the water never leave the trust, and the water never leaves the district. The trust is responsible for the upkeep of the irrigation infrastructure and charges the irrigators accordingly. The farmers may compete over what they grow, how they grow it and how they market it, but the trust ties them together in co-dependence. At least, that's the way it has worked right up until this very week. Now, as I sit ignoring my alleged fish, change has arrived. Not exactly the crumbling of the Berlin Wall, but change nevertheless. New laws have been gazetted that change the rules of the trust. Farmers will now be able to convert their right to water into an individual water licence, which means they can buy and sell permanent water licences separate from land titles, including selling water out of the trust district. Eventually, the trust will devolve its collective water licence into separate farmer-held licences. The collective effect is to add a dose of market forces to a system that has been characterised by central planning. The idea, at least in theory, is to allow water to find its own level, not by gravity but by what the market is willing to pay, thereby ensuring it's put to the best use. Soon, the Renmark Hotel may be all that's left of the town's old collectivism.

I wait under a cloudless sky at Tim Whetstone's packing sheds. Canberra was decidedly autumnal when I left on this latest trip, but down here in the Riverland the summer has lingered into late April. It's nothing compared to January, of course, when the air temperature here climbed above 48 degrees. And that was in the shade. Out in direct sunlight it was hotter. Fruit boiled inside its own skin, and trees denied water even for an hour began to wither. Even now, with the temperature climbing through the high twenties, I can feel the solar weight on my fair skin, and I apply another layer of SPF30+. Tim rolls up in a hefty four-wheel drive and we talk as we wheel around his orange grove. It's a very different tour than Orange World, and Tim has little of Mario's infectious bustle.

All the water licences along the South Australian Murray are high-security licences, just like Mario's in New South Wales. But in the Riverland the words have become a mockery. The high-security water is meant to have 97 per cent reliability, meaning farmers should get their full allocations in ninety-seven years out of 100. With this in mind, Riverland farmers have planted permanent crops—grapes and citrus, almonds and olives—together with all the necessary infrastructure to support them. The high level of reliability is underwritten by an interstate agreement guaranteeing South Australia a set amount of Murray River water: 1850 billion litres. But when there's not enough water, there's not enough water, and the state's overall allocation is cut. With no significant tributaries of its own flowing into the Murray, South Australia is totally dependent on what comes downriver. Tim tells me the secure world of the Riverland irrigators began caving in three summers ago, when their high-security allocations were cut to 60 per cent. Last season they were 32 per cent. This year they are at 18 per cent. Next season, less than three months away, looks even worse. It's a disaster. The choice for Riverland farmers is stark: buy water or let your plants die.

To Tim, the entire system is fundamentally unfair. High-security licence holders on the New South Wales Murray and Murrumbidgee are on 95 per cent of their allocations. But what really rankles is the water going to low-security irrigators in New South Wales: the broadacre irrigators. On the Murray, at places like Wakool, the allocation is only 9 per cent, but on the Murrumbidgee it's 21 per cent and on the Darling below Menindee it's 50 per cent. Tim questions how it can be that low-security farmers upriver can get more water than high-security irrigators in the Riverland.

Tim pulls up next to a row of young trees. He explains how he's progressively cutting back or cutting out older, less-productive trees, planting new varieties, installing more-efficient micro-irrigation systems. There was a time when he watered the trees once a week, in the evenings when evaporation was low, using sprinklers that threw water 4 or 5 metres. Now Tim waters daily, during the heat of the day when the trees are thirsty, the water rationed out to individual trees. He can't afford the next step: subsurface drip-irrigation, regulated by computerised probes that keep a constant supply going to the trees. Over on a neighbouring property, I see a sprinkler chugging out water in a wide and wasteful circle. Tim says individual growers have their own opinions on the cost-effectiveness of putting in new infrastructure. Which may simply be a polite way of saying that not all of them can afford it.

'The water restrictions are hurting the growers financially, but we're also dealing with the commodity prices', says Tim. 'If we were on good commodity prices, we could bite the bullet and go out and lease water. But a lot of people are getting to the point where, this year in particular, with wine grape prices very low, they can't afford to lease water. So we are seeing a lot of trees cut back or pulled out and vines left to die.' Tim says growers are operating on negative cash flows. 'It's been happening for a couple of years now. It's beginning to hurt big time. Banks are starting to pull on properties and saying, "Sorry, no more borrowings".'

We continue moving through the grove, startling a couple of emus that have wandered in from the desert. They're a reminder that just beyond this bright-green construction, the desert lies waiting. We climb a small hill. There are no canals to be seen. This is another gripe in the Riverland. The entire Renmark Irrigation System was converted to sealed pipes back in the early 1970s; there's no loss to evaporation or seepage or dodgy Dethridge wheels. The shortages of recent years have caused another wave of change as hard-pressed growers move to drip irrigation. The same evolutionary pressures that have been pressing down on Mallee farmers for decades have started asserting themselves on the Riverland. Of course, evolution is a fine thing for the winners; it loses its gloss among dinosaurs and dodos. Evolution in agriculture requires time or, failing that, money. On the Riverland there's not much left of either. Tim is doing what he can, but he doesn't have the money or the economies of scale for the really high-tech options. And with uncertainty over commodity prices, even those who have the money must wonder if it would be better invested outside agriculture.

Tim's a handsome bloke in his late forties, with a strong jawline and a movie-star chin. When the drought started, he must have thought himself bulletproof. But he's been forced to sell his real-estate investments, and now he's selling his vineyard, all in order to keep the citrus grove afloat. And he's lost more than that. I ask him if his wife works off the farm. He tells me she's left him, that the financial strain helped end his marriage.

'We'd gone along reasonably comfortably for the majority of time we were here together and all of a sudden there's not the cash reserve there. The lifestyle disappears. It all adds to it', he says, looking straight ahead, concentrating on driving. 'We were together for thirty-five years. And she's been gone twelve months.' His eyes are hidden behind his sunglasses; for a moment the granite jaw quivers. Then Tim is asserting he's relatively well-off, and explains why. At nearby Loxton, a major wine maker recently reneged on wine contracts. Growers who had mortgaged themselves to lease water, who had fruit on their vines, suddenly found themselves without a buyer. Other wine companies followed suit. Rumours circulated that some wineries were reneging on contracts, then going next door and offering half the price.

Tim reckons something like 30 per cent of wine growers around Loxton are on antidepressants. 'The suicide rate was right up over there. The authorities don't want to give out the numbers. It's happening more and more. They're overrun by uncertainty; they're overrun by something they've worked all their life for disappearing. And they sit there and say "I don't see a future" … It's a terrible world out there, people suiciding, people walking into the bank and throwing the keys on the desk and saying "I can't do it anymore. I'm out of here".'

We arrive back at the packing shed. Tim is cracking hardy, saying there is a future here. Rather sheepishly, he tells me he intends running for the state parliament; he'll be up against the local member, the water minister. I tell him he must be a sucker for punishment. The sun is growing hotter as we shake hands in farewell. I ask if he reckons rain is on the way, as the radio has been reporting for days now.

'Absolutely. Absolutely. I can feel it. You get a sense for it. We might even get some this afternoon', Tim says.

'There's not a cloud in the sky', I say.

'Yes there is. Look over there.'

I follow his pointing hand. Sitting just above the horizon is the smallest powder-puff cloud imaginable.

'See', he says. 'We're saved.'

⌒

South Australia's sieve-like guarantee of river water results from negotiations that began decades before Federation and still continue today. South Australia, Victoria and New South Wales met as early as 1863 to discuss the development of the river system. In those days, the priority was river navigation. Discussion revolved around building weirs and locks to sustain steamer traffic through the low rivers of summer. But irrigation soon emerged as a competing interest. The colonies knew cooperation was necessary but jealously guarded their sovereignty nevertheless, with customs houses erected where borders and rivers intersected. Before agreeing to federation, they insisted their riparian rights be safeguarded in section 100 of the Constitution: 'The Commonwealth shall not, by any law or regulation of trade or commerce, abridge the rights of the State or of the residents therein to reasonable use of the waters of the rivers for conservation or irrigation'. (In the language of the day, 'conservation' meant conserving water in dams, not preserving the environment.) Any negotiation on water sharing was the prerogative of the states, with the new federal government left as a coordinator at best, an interested observer at worst. Section 100 of the Constitution remains unchanged.

New South Wales had the whip hand. It controlled the Murray. Its border with Victoria started at the high-water mark on the southern bank. In turn, this gave Victoria an incentive to exploit every drop of water in its inland rivers before they ran into the Murray, and thereby into New South Wales. For its part, New South Wales had no legal obligation to let water flow through to South Australia. It took until 1914 to reach a formal agreement: the River Murray Waters Agreement. It gave the Commonwealth a coordinating role, but the states retained the real power. The agreement gave birth to the modern river; it led to the building of Hume Dam to guarantee river flow and the construction of the inland sea. It also set out the formula for sharing water between the states. Its legacy is still very much with us today, for once a right to water is bestowed, it's difficult and expensive to take it away again. Just ask Penny Wong.

The agreement gave South Australia a guaranteed amount of water: 1850 billion litres. New South Wales and Victoria share the remaining water in the Murray system fifty-fifty. But the amount flowing to South Australia can be reduced at times of extreme water shortage, according to a predetermined formula. That's why Tim Whetstone's high-security allocation is no longer high-security. Importantly, under the agreement, the source of the water was clearly defined: the Murray River system, not the Murray–Darling Basin as a whole. The Murray River system covers the alpine catchment east of Hume Dam, including Dartmouth. It then covers the main channel of the Murray from Albury to the sea. It also includes the lower Darling River, but only after the levels in Menindee are high enough. Significantly, aside from equivocal coverage of the lower Darling, the agreement does not cover any of the Murray's major tributaries. The *Water Amendment Act 2008* states: 'New South Wales and Victoria are each entitled to use: (a) all the water in tributaries of the upper River Murray downstream of Doctors Point (Albury) within its territory, before it reaches the River Murray'.

In a non-drought year, the combined flow of the Murrumbidgee and Victoria's Goulburn River is close to one and a half times that of the Murray. In such times, more than 80 per cent of the Murray's water is committed, compared to about half of the Murrumbidgee's and just one-third of the Goulburn's. So when drought comes, the Murray quickly runs short of water. Which explains a lot. It helps explain why low-security farmers at Wakool, fed by the Murray, are on 9 per cent allocations while low-security irrigators on the Murrumbidgee are on 21 per cent. It helps explain why the Murray's mountain dams are 14 per cent full while the Murrumbidgee's are 34 per cent full. But most of all it explains why there is bugger-all water flowing from the Murrumbidgee into the Murray.

Returning to Renmark from Tim Whetstone's place, I stop at a roadside winery. Perhaps a few bottles will cheer me up. The cellar door is clean and purpose-built, with tiles on the floor and the faint smell of disinfectant. A wooden bar displaying assorted bottles and glasses extends along one wall, supporting a couple of grey nomads who are swishing wine appreciatively around their glasses. There's an elderly woman standing behind the bar talking to them. The sound of heavy rock, Cream playing 'Sunshine of Your Love', is throbbing out of overhead speakers. 'DAH-DA DAH-DAH. DUM DUM DUM DE-DUM DUM' pulses Eric Clapton's opening riff. I catch the old woman's attention with a wave.

Her name is Helen. She was born in Greece, lived in Melbourne when she first arrived in Australia, then came to Renmark forty years ago with her husband to grow grapes. Ten years ago they started making their own wine, with their son as winemaker. I sample a wine I've never heard of: tempranillo. Helen says it's a Spanish grape. It swirls a bright crimson around the glass and smells a touch acidic. I take a tentative taste as Cream starts playing 'Strange Brew'. They're not wrong. I quickly swallow before it can take the enamel off my teeth. 'Very nice', I say. Egged on by her smile and the 1960s supergroup, I try another mystery variety. Helen is already pouring it when she tells me it's the family's own rendering of retsina. I take a break from wine sampling and instead ask Helen how the winery is faring. She says it's been a tough couple of years. Last year, after the district's irrigation allocation was cut, the family ended up paying $50 000 to lease water. This year, they've read the market better, spending about $20 000. But the heatwave has cut their crop almost in half. And, like everyone else, the price they're receiving for their grapes has plummeted.

'We must sell for more than $500 a tonne', says Helen. 'If they buy for $150, we lose money. That's why we started making wine. But there's not much profit. Not much. We make less than $20 000 profit this year.' Cream, who appear to be eavesdropping on our conversation, have moved onto 'I Went Down to the Crossroads'. Helen seems oblivious to the music; her son must have it on a loop. I try some more wine: the 2005 and 2006 Rieslings. Much more acceptable. Well, drinkable. They're $11 a bottle so I buy half a dozen. As I leave, Jack Bruce is singing, 'If it wasn't for bad luck, I wouldn't have no luck at all ...'

Back in town, I wander into Renmark International Backpackers. The hostel offers not just accommodation, but help with visas and finding work picking fruit. I wonder how the backpackers are doing, with grape prices plummeting and more and more vines harvested mechanically. There's no-one at the hostel counter, just a note with a phone number. I try yelling instead. No answer, but I can hear some activity so I wander through. Inside a communal kitchen, a young man in a black T-shirt is eating toast, and three pretty Asian girls are chatting among themselves. I ask where the owners are, or if there is someone in charge. 'No', says the young man. In the peculiar hybrid accent of the French Canadian, he tells me they've gone to Adelaide and won't be back until Sunday; today is Thursday. I tell him I'm writing a book. He says I should have come the previous evening: the owners threw a big going-away party for a couple of long-term residents.

The youth tells me his name is Martin Decoster, that he's twenty-one, and that he's come to Renmark to pick grapes. But there's not much work around, and what there is doesn't pay well. He tells me he was earning about $60 for a seven-hour day at one place, while a friend was earning closer to $150 doing the same work down the road. 'You know, I'm not so slow', says Martin. But he likes Renmark. The weather's good and the hostel is right in town; the backpackers can walk down to the hotel or the Renmark Club and drink with the locals.

The girls are named Fiona, Tiya and Vivian, their English names. They're from Taiwan. Their English is limited, but they tell me they're in Australia for a year on a working holiday. They've been in Renmark for two weeks, waiting unsuccessfully for work.

'You like it here?' I ask.

'Oh yes', says Fiona.

'What's good?'

'Wine', she says with a giggle.

'Seriously?'

'Oh yes.'

In their broken English they say that wine is expensive in Taiwan, that everyone drinks beer and spirits. Now they've found it's possible to buy good local wine for $5 a bottle. 'Five', Fiona says, holding up her fingers for emphasis. 'No sign.'

'No sign? Um … no label?'

'Yes. Yes. No label sign.'

'A cleanskin. It's called a cleanskin.'

'Cleanskin', they repeat, giggling as they practise the new word. 'Cleanskin is very good', beams Fiona while her friends chortle away. I guess it's an ill wind.

⌒

I drive down to lock five by crossing back over the bridge to Paringa. Alongside the river, a flotilla of houseboats is moored, looking more like a row of flats than anything designed to float. There's *Jabiru*, *Firefly*, *Pure Magic* and *Aqua Dreaming*. Lock five is called a lock, but like the other barriers on the inland sea, it's actually a weir, with the lock over on one side. There's a modicum of water flowing over the weir's gates, but above and below the structure, the river is wide and flat with little perceptible flow. A helpful sign indicates the flow is actually some 3 gigalitres a day, the absolute minimum required to meet environmental flows. That's the deceptive thing about the inland sea. The lock pools are always full or near full. Unlike the flood plain, there's no way of telling how healthy the main channel of the river is by simply looking at it, not unless it's choked by duck weed or blue-green algae. So I've developed my own measure. Back at the Renmark Hotel I drink a glass of water. Up at Mildura the water had a detectable taint, a soft insidiousness mixed with salt. Not the best, but potable. Eau de Renmark is noticeably worse. There is something dank in it, something unpleasant, something the pool chemical taste can't fully disguise. I swallow a mouthful and then do what I should have done back at the winery: spit the rest out.

I eat dinner at the Renmark Club, another bolshevist venture just along the river from the hotel. It's a big place, like a football or bowls club, bearing few similarities to the eponymous anachronism in Echuca. The club has a large deck overlooking the river. I have it pretty much to myself, with only the occasional smoker drifting out to satisfy their craving. Why anyone would stay inside on an evening like this is beyond me. The air is warm, with a gentle breeze. There is no sign of Tim Whetstone's rain. I don't know whether it's the Chaffeys' original utopian vision or the communal tendencies of their heirs, but Renmark has managed to preserve public access to the riverbank. On the town side you can walk for miles alongside the Murray, admiring the old houses, looking at the houseboats and the occasional waterskier. The far bank is uninterrupted bushland, the contagious promise of wilderness. The club has mounted two powerful spotlights on its roof, lighting the river and the distant red gums; the water is a milky green under the lights, and a couple of ducks paddle about enjoying an evening feed. Despite the proximity of water, I've learnt my lesson about ordering fish. Besides, it's Thursday, 'steak and grape night': a thick T-bone and a glass of local wine for next to nothing. And as I eat, enjoying the soft river evening, I wonder if reform can be instituted in time, if the latest federally brokered agreements can achieve much.

I pull out my map of the Murray-Darling Basin, now frayed and travel-worn, and do something I should have thought of weeks ago. I turn it upside down. Viewed the normal way up, with north at the top, the natural flow of the river seems self-evident: water runs effortlessly down the Darling from Queensland and northern New South Wales towards South Australia. By contrast, the water coming along the Murrumbidgee, Murray and Goulburn Rivers appears as if it needs to almost climb uphill to get to Mildura and beyond. But it's an illusion. Bourke, where the Darling gathers itself, is just 106 metres above sea level. On the Murray system, Dartmouth is at 486 metres and Hume at 192 metres; on the Murrumbidgee system, Blowering Dam is at 379 metres and Burrinjuck at 361 metres; on the Goulburn River, Eildon Dam is at 230 metres. The upside-down map is a better reflection of reality. Try it. Now the sparse water of the Darling struggles to climb the long slope to South Australia, while the waters of the basin's three most plenteous rivers have a relatively short and easy journey. If South Australians want water to revitalise the Riverland and replenish the lower lakes, they shouldn't obsess about Queensland cotton farms and the evaporation rates at Menindee. They should turn their maps upside down and look towards the mountains. But it still doesn't solve the problem, for South Australians can expect no greater share of the Murray and very little from its tributaries.

John Howard had a serious crack at reform in January 2007, more than ten years into his stint in office. Either the crisis in the basin had grown so severe that he was willing to take on his National Party rump, or he was desperate for an issue to deflect Kevin Rudd's relentless push towards the Lodge. Maybe both. On the eve of Australia Day, the prime minister wrapped himself in the flag, declared himself a nationalist, and announced a plan to have the states cede to the Commonwealth their century-old constitutional power over the rivers. He used the time-honoured method, a huge pile of cash, announcing a $10 billion plan. Was it Paul Keating who said, 'Don't stand between a premier and a bucket of money'?

But in this case it didn't work. The Victorian Government resisted the temptation. Steve Bracks had caused a major upset in the 1999 state election when his Labor Party wrested power from a cocky Jeff Kennett. Labor won that election in the bush, campaigning hard to reinforce the perception that Kennett was Melbourne-centric, that he had neglected the rest of the state. Perhaps Premier Bracks feared that agreeing to Howard's plan would see him lose power the same way he'd won it.

A year later, both Howard and Bracks were gone—Howard put to the sword and Bracks falling on his. Kevin Rudd's new government was eager to demonstrate that it could achieve in its first year what Howard had been unable to accomplish in eleven. It left the money on the table, indeed raising the pot to $12.9 billion, as the premiers salivated. But the new Victorian leader, John Brumby, had helped devise the rural strategy that had felled Kennett; he was as reluctant as Bracks had been to surrender any Victorian water. But eventually even he couldn't resist. How could he? The deal gave him money and it gave him water. Indeed, it hardly warrants the description; this wasn't a deal so much as a gift. On top of the money already promised to the states, Brumby secured an extra $1 billion of federal money to upgrade Victoria's irrigation infrastructure. A billion dollars is a lot of money. A billion dollars to line canals, to replace open ditches with pipes, to replace Dethridge wheels with computerised flumes. A billion dollars to spend in marginal electorates, creating jobs and securing votes. In return, half the water saved, but only half, would flow into the Murray. So Victorian irrigators could get more water not less, paid for by the federal government, while Brumby takes the credit. How good a deal is that? But wait; there's more.

The federal government has won the authority to mandate environmental flows and to set limits on how much water can be diverted for agriculture across the basin. That's a major achievement. But it doesn't have the power to give South Australian irrigators more water; both New South Wales and Victoria retain the right to veto any changes to how the states share the Murray's waters. So John Brumby could sign the deal confident that his state's irrigators would still fare better in times of drought than those in South Australia. But the real kicker, the element that must have sealed it for Brumby, the one that eliminated any ambiguity, was that under the new plan, existing state water resource schemes remain unaltered. Victoria's plan extends until 2019. Little wonder Brumby signed; by 2019 it sure won't be his problem. And so Rudd and Wong got their deal when Howard didn't. Terrific. We're getting a shiny new Murray-Darling Basin Authority; such a pity it won't have much authority over the Murray-Darling Basin for the first decade of its existence.

Maybe I'm being too pessimistic: seeing a river half-empty instead of a river half-full. The deal will deliver some additional water. Canberra is spending over $3 billion buying permanent rights from farmers and returning the water to the river. It will invest billions more in improving irrigation infrastructure. And the plan will designate greater environmental flows. It all helps. But then again, sitting in South Australia as I am, it's hard to see it helping enough. Not if climate change is already beginning to bite.

I wonder what Victorians think of this. As Victorians, they must be cheering Brumby for extracting the best possible deal for their state. That, after all, is his job. But as Australians, perhaps they look at South Australia's lower lakes and recognise that 2019 is a very long way off.

Rameses II did not rule a divided land. The Nile did not separate one state from another. He was a god, not to be swayed by the lobbying of obdurate farmers, or bothered by the campaigns of recalcitrant greenies, or influenced by the manoeuvrings of intractable premiers. We're stuck with mere mortals. I don't hold out much hope that in the millennia to come, larger-than-life statues of any present-day politician will be propped up outside motels to attract passing trade. I look out across the river, and in my mind's eye I see prosperous Renmark beginning to fade, just as Murrayville has faded—just as Bourke and Wakool have faded.

9

JOURNEY'S END

The receding shoreline of Lake Alexandrina

Promised for a week, the rain arrives as I drive along the Murray westwards from Renmark. With no significant falls since before Christmas, the approaching front has taken on the nature of prophecy, dividing the community between believers and sceptics. Now it comes. Not a clearly defined front, no storm clouds advancing en masse like that time near Bourke, but rather a gradual build-up—the light dimming, the temperature dropping, as the first high-altitude veiling shades the sun. A tentative raindrop or two falls, as if scouting unknown territory. Under the diffuse light, the road takes me away from the river's verdure, into the arid lands above Goyder's Line.

South Australian Surveyor-General George Goyder mapped his line in 1865. It divides the state's agriculture in two: south of the line, cropping is viable; north of the line, the land is suitable only for grazing. Another of those invisible lines. Goyder ignored erratic rainfall patterns, inadequate statistics and ill-informed opinion, and instead looked to the land to be his guide. He drew his line along the borders of native vegetation patterns, placing mallee scrub to the south and saltbush to the north. For almost a century and a half, his delineation has proven remarkably prescient. But now, scientists are warning that climate change is likely to push the line further south. So I'm surprised when I see evidence of cropping, hay stubble scattered flat on the red earth of a roadside paddock, with a gaggle of emus scrounging for tidbits. But I'm mistaken; the hay is not the remains of a crop. A grazier has been hand feeding. The sheep have gone now, perhaps forever, leaving the paddock utterly denuded, eaten down to the red earth. Just more dust to feed the great red haze spied from Conroy's Gap.

Back on the river, at the Overland Corner Hotel, sitting by itself down on the flood plain, a line two-thirds up a stone wall shows the level of the 1956 floods, the highest since the hotel was built in 1860. I walk through the beer garden and climb up the levee. The river is beyond sight, hundreds of metres away across the flood plain. A few more raindrops fall, harbingers, and I return to the car. Past Overland Corner the river is more and more contained by cliffs and escarpments. It's not a narrow gorge. Rather, the cliffs act as boundaries, with the river allowed to amble around the floor of the valley, hugging a cliff here, meandering across the silt plain there. I cross the river near Cadell, a uniquely South Australian experience. I inch down to boom gates where I wait for a ferry to collect me. It approaches, guided by steel cables. Why a ferry? The banks are steep here, the river relatively narrow: a natural site for a bridge. The vessel provides its own answer as it glides towards me, barely pulling on the cables to correct its passage: there is negligible current in the inland sea. Upriver, ferry crossings could be made precarious by releases from Hume Dam, sudden surges from rain-fuelled tributaries, floating logs. Not here, where crossing the river is like crossing a millpond. The ferry itself is little more than a guided pontoon, its deck sitting just a metre or so above the surface.

Past the ferry, sheer grey-white limestone walls drop 30 metres to the river. After all those weeks on the feckless plain, the river has found some purpose. The cliffs swing it abruptly from its westward meander and direct it south instead, towards the sea. I follow the river along its eastern edge. The limestone cliff on the far shore diminishes, while another, much higher, climbs on my side: towering red rock. I stop by the cliff edge. There are no safety rails. I shuffle timorously to the precipice, drop a pebble, listen for the splash. One, two, three, four, plop. It must be 70 metres down. The far side of the valley is a kilometre or two away, with a low flood plain between the escarpment and the river. Along the opposite bank, a community of holiday houses and fishing huts has burgeoned. Even in the rain they look idyllic. I can picture myself lounging on a deck of a clement summer evening, the bushland on three sides, the river before me, and soaring above, the face of the cliff, lit auburn by the setting sun. Somewhere adjacent to paradise. But another thought intrudes. All the houses look relatively new, and they must be. For if a flood were to come through this valley, the water would have nowhere to go; there is no boundless plain to dissipate the energy. The low-lying weirs of the inland sea would do nothing to impede such a flood. Indeed, they are designed to be removed altogether at such times. The holiday houses would be swept away. Just because it's been decades since a major flood doesn't mean another one is impossible, climate change or no climate change. Even if the weather grows hotter and drier, it may also become more volatile. As if to echo my concerns, the wind gusts and a squall of rain comes up the valley, scrubbing the river from glass to brushed steel. I feel no desire to dance in the rain amid the deepening greyness.

I reach Blanchetown, home to lock one, the first weir on the Murray. The weir is 274 kilometres from the river mouth, yet the surface of its pool sits just 3 metres above sea level. Below the weir, the water on this autumn day is half a metre *below* sea level. The river has run out of impetus: impetus and water. From here to the sea, it doesn't flow. It just sits there, the water that spills over the weir topping it up like a leaky bathtub rather than creating a genuine flow. Call it an inland sea, or a lake, or a canal: all are more appropriate descriptions than 'river'. The grey day grows greyer.

Blanchetown was founded in 1855 on the cliffs above the river, replacing an earlier settlement, Moorundie, sited a few kilometres to the south. Moorundie, which once made the familiar claim to be the earliest settlement on the Murray, was built on river flats and consequently destroyed by floods. Blanchetown has forgotten the lesson, spilling down the side of the cliff as if falling through the bottom of a sodden bag. Atop the escarpment is a handsome sandstone post office, and a little further down is an adequate looking pub, but the houses at the bottom of the slope are plain and unkempt. Near the river, an Aussie Rules oval lies neglected: nothing but bare earth and awry goalposts. Next to the oval is a concrete-brick shed with 'Blanchetown Golf Club' painted in rough letters on its side. There's a thin slot of rusted metal in the wall, like a postbox: '$3.00. Green fees here please'. A larger sign states 'NO BBQS ALLOWED'. But the signs are faded and there's a sense of defeat. I search for the golf course, but find only bushland. There is no first tee, no flag beckoning from a distant green, nothing to separate fairway from rough, only the straggly bush. On the corrugated-iron fence of the nearest house, someone has sprayed 'WAT A SHITHOLE!' in large black letters. The rain has released a smell from the flood plain, an odour of marshes and mudflats.

I've been in town for half an hour and I'm yet to see a single person. There's a sandwich-board sign beside the road leading back up the escarpment: 'General Store. Open 6 Days'. But when I get to the store, a handwritten sign on the door says 'Enjoying some R&R. See you end of May'. It's the middle of April. The job centre next door is open for business, but I'm looking for lunch, not employment.

I find another store that sells takeaway food. I sit outside under an awning, watching the rain intensify as I pick my way through a steak sandwich that's more gristle than meat. The taciturn couple behind the counter and a truck driver intent on a hamburger and chips are the only signs of life. I visit the lock, but it's closed to the public for construction work and looks deserted. Through the cyclone-wire fence and the drizzle I can see a thin film of water spilling over the weir, with a crowd of cormorants drifting silently on the turbulence. Across the road I find a bubbler. A sign informs me the Blanchetown water supply is a $1 million project funded by the federal, state and local governments. The pressure in the bubbler is scant but enough to get a taste. I spit it back out. It's worse than at Renmark.

The rain is washing in more consistently now. The greyness is no longer restricted to the clouds; it has floated downwards, entering the landscape and wrapping Blanchetown in its tendrils. Time to leave. Further down the river, this river where even the meaning of the word 'down' is beginning to fray, beyond Swan Reach, the clouds lift one last time and I spy a road descending the cliff to the water. At the bottom, there's a nondescript picnic area. I walk along the base of the cliff, then scramble up a few metres. The rock is pockmarked, eroded by rain, wind and the occasional flood. I fossick around and find the fossilised remains of shells. They've lain here for an eternity, buried in the solidified bed of an ancient shallow sea only to be exposed again by a whim of geology. The sea is long gone but it has bequeathed a pervasive legacy: salt. Salt is the great enemy of the Murray River, and the further the river travels, the more aggressive the salt becomes. It's predicted that within a decade, Adelaide's drinking water, extracted far upstream from Blanchetown and Swan Reach, will exceed the salinity limits set by the World Health Organization 40 per cent of the time. The South Australian capital is not a place to have high blood pressure.

When I was in Mildura, I'd visited a salt-reclamation scheme, part of a hidden infrastructure, another engineering solution struggling to keep the river viable. It was half a week ago, but now, standing in the damp grey under the cliff, it seems like half a century. Mildura had been almost hot, sitting under a pale blue dome with temperatures in the high twenties, so different from this drizzle-sodden day. Away from the river, out at the Mourquong salt mine, the temperature had pushed well into the thirties as the sun bounced off a snow-white field of drying salt. Even through my polarising sunglasses, the effect had been dazzling and disorientating. Despite the desert heat, I felt as if I were at the snowfields, preparing to put on cross-country skis and head off across the glaring expanse. Mine production manager Leighton Schmidt told me the January heat wave had been great for evaporation and salt production, but the workers had been forced to down tools by mid-morning. In town the temperature had hit 48 degrees; Leighton reckons on the salt field it had climbed to 55 degrees. The world record of nearly 58 degrees was set in Libya in 1922. Just the thought had me applying more sun cream and taking another slug of water.

When the primordial sea withdrew, the salt it left behind gradually leached into the water table. The semi-arid climate kept the water table below the surface, as native vegetation soaked up most of the rain that did fall. But irrigators cleared the deep-rooted mallee scrub and flooded the land with water. The water table rose, bringing salt to the surface. I'd suggested to Leighton that therefore the drought must be good for reducing salinity, because less irrigation must lead to a fall in the water table. Leighton had listened politely, told me I was half-right, and then politely explained why I was wrong. Running 15 or 20 metres under the basin is a massive aquifer. The further west it travels, the more saline it becomes. The problem emerges where the aquifer intersects the rivers. When the rivers are full, the fresh water pushes outwards against the banks, maintaining a pressure that prevents the saline aquifers from flowing into the river. But if the river is low, there is no outwards pressure, and plumes of brine seep into the Murray. Over the decades, irrigation has increased the pressure within the aquifer, with long-term effects.

So an engineering solution has been developed. Bores have been sunk near the rivers to intercept the aquifers and relieve the pressure. The brine is pumped out into evaporation pans where the salt is left quarantined on the surface. At Mourquong, the brine from the government storage ponds is diverted into the private evaporation pans of the mine.

'It's an annual process', Leighton had explained. 'The brine starts going in … in September, and then all the way through to March we keep topping it up as the sun takes it off. And as of March, then we pump what's left off. That's what we've been doing the last couple of weeks. Now we're starting to harvest … We get about 25 000 tonnes a year. We make industrial salt, hide salt, agri-salt and food-grade salt.' Leighton told me that the food-grade salt is pink and the owners market it as a gourmet product. 'It won an award for best salt in the world or something in California. It has a softer taste.' For a mine production manager, Leighton made a fair fist of being a marketer.

Another squall comes carving up the river, bringing me back to the present. I scramble down from the cliff and race for the car. I cross the river on another ferry at Mannum and head southwards again along the western bank. As night turns the grey to black, the rain becomes heavier, more constant. The periods between squalls grow less and less until there are no breaks at all, only a continuous batter-ing of the windscreen, challenging the capacity of the metronomic wipers. I reach Murray Bridge where, out of the gloom, an apparition appears. It's a life-sized fibreglass elephant, sitting atop an irrigation and fodder supply shop. It's frozen in mid-trumpet, head uplifted in the rain, joyous in the downpour. But the elephant is white. What a strange yet appropriate mascot for an irrigation supply business. It joins Rameses II and the others in the pantheon. A large pub emerges from the rain, glowing in welcome, and I scurry inside to find a room. But I'm mistaken: the glow is not the warmth of hospitality, just the cold gaudiness of poker machines. The entire ground floor has been gutted to make way for a 'gaming parlour' encircled by a thin *cordon sanitaire* of bars. The front bar, squashed into a long corridor a few metres wide, offers little respite: one end is dominated by PubTAB, with television screens displaying interstate greyhound and harness racing; the other is crowded with Keno screens. There's only one customer, laboriously filling out a betting slip. But there are cheap rooms upstairs and I take one gratefully. I taste the water. It's undrinkable. Leighton and his colleagues may be keeping a lot of salt out of the river, but other, more elusive substances are leaving their mark. I brave the driving rain to retrieve a bottle of mineral water from the car.

In the morning I descend to breakfast. To gain access to the one semi-functional part of the hotel, the bistro, I'm forced to traverse the gaming parlour. Inside, there are no windows and no clocks; time has stood still. Outside it's eight in the morning; inside it's midnight. Four or five women and a couple of men remain at their machines, attached by the umbilical of compulsion, faces devoid of emotion, eyes transfixed, their hands flicking occasionally at a button. The glowing machines, with their Aztec kings, red Indians and Asian coquettes, seem more alive than the people—the succubi of our times. Dante-like, I pass the unseeing, the doomed and the tormented, not quite breathing until I gain the sanctuary of the bistro. In the *Divine Comedy*, Dante placed Purgatory in the Southern Hemisphere. He wrote that when it was sunset in Jerusalem, it was sunrise in Purgatory. He didn't mention Murray Bridge by name.

The bistro is hosting a power breakfast. A local real-estate company is handing out monthly achievement awards over Nescafé and toast. Their unnatural early-morning cheerfulness is almost as unsettling as the living dead in the next room. There are no other guests and no sign of the promised buffet breakfast. I nab a passing barman.

'Can I help you?' he asks politely.

'I'm after breakfast. I stayed here last night. Upstairs.'

'No problem. Grab what you like from under the counter over there. Anything you can't find, just swipe it from those guys', he says, tilting his head towards the real-estate people. 'And there's tons of free coffee in the gaming parlour.'

His cheery disposition puts him firmly on the right side of the Styx. I'm feeling somewhere mid-stream myself, not yet having ingested my first coffee of the day.

'Anything else?' he asks.

'Yeah. What's that smell?'

'Smell? Oh that. That's the dog food factory. You get used to it.' And he beams a fresh-faced smile. It's going to be one of those days.

Two hours and two espressos later, I'm in a much better mood. I'm sitting by the shores of Lake Alexandrina, the larger of the two lower lakes that lie at the end of the Murray. The rain that pelted down throughout much of the morning has let up, the wind has died, and the sun has broken through. Next to me is Clem Mason, dairy farmer and champion of the lakes. We sit on a bench by his boathouse, looking out across his private beach. It would be an idyllic scene, except the lake no longer laps the shoreline. It's retreated 150 metres from Clem's house, leaving behind sand, mud and sporadic growth. It glistens in the distance, too substantial to be a mirage but, according to Clem, illusory nevertheless.

'This year we had no mosquitoes, no frogs, no snakes', he says. 'This is an ecosystem breaking down. We sit and look out there now and we say, "The water is a long way out, but it's still there. Perhaps we're not so bad off. Perhaps our system can survive". But the quality of that water is breaking down. We are going to be the first Western country who is going to have to put our hands up and say, "We've failed to protect a wetland listed by the world as being of immense importance. We are now too late". We're coming up with all sorts of bandaid solutions to save the last bastions of migrating birds, of fish. It's vandalism on a huge scale.'

Clem Mason is one of the last dairy farmers. When I arrive at his lakeshore property, Masondrina, he's in the yards, sorting cattle. Nearby, large spirals of hay are sitting ready for transporting. The property has received a couple of centimetres of rain overnight, and more is on its way. But down by the shore, looking at the lake, Clem tells me of how dire the situation has become. Not for his property or his livelihood—Masondrina is large enough and sufficiently close to the sea to benefit from coastal rainfall—but for the lakes. He launches into a stream-of-consciousness monologue. It takes me fifteen minutes to interpose my first question. By then his voice has taken on the cadence of a preacher, his sentences broken down into individual emphatic phrases.

'I've been here for twenty-five years ... It's two years since we irrigated out of the lake ... We stopped irrigating because we couldn't access it ... And the quality is dropping ... It's more than 100 metres out, but I could walk another 300 metres out past there ... and it would come up to my shins ... If we get a wind that pushes it away ... I could go another 500 metres out.'

The Murray River finally runs out of elevation up near Blanchetown, yet it stretches for another 200 kilometres, constrained by the banks and levees built up during more plenteous years, the years when it still flowed, until it reaches Wellington. At Wellington, defeated by the flatness of the land, the water spreads out into Lake Alexandrina, a massive freshwater body. As the water in Lake Alexandrina builds up, it flows through a narrow channel into a smaller lake lying to the east, Lake Albert. When enough water comes down the Murray, more than the lakes lose to evaporation and, in previous years, to irrigation, the water rises above sea level and spills through the barrages into the Coorong. The barrages are like long low walls, with roadways on top and water gates underneath. There are five, ranging in length between 630 metres and 3.6 kilometres, separating the lower lakes from the Coorong. The Coorong itself is a unique aquatic environment, a narrow lagoon stretching 140 kilometres down the coast, separated from the ocean by a thin range of sand dunes. The barrages and the lakes lie at its northern end, close to where the Coorong connects to Encounter Bay through a narrow opening; it's still known as the mouth of the Murray River, although its connection with the river is becoming increasingly tenuous. The Coorong is a shifting mixture of fresh water from the lakes, sea water moving back and forth through the mouth, and hyper-saline water in its southern reaches, where evaporation and lack of freshwater inflows have conspired to turn the water up to seven times as salty as sea water.

Now this system, in place since the barrages were erected in the 1930s, is failing. Not enough water has been coming down the Murray because of drought and overallocation. Fresh water has not flowed through the barrages for three years. The level of the lakes has fallen to a metre or more below sea level. Only the barrages hold the sea water back. Irrigation from the lakes has stopped, the water too low to access even as its quality deteriorates. Worse, the exposed shorelines threaten to turn irreversibly toxic. Decomposing organic material, built up over millennia, is buried in the mud of the lakebeds. While it remains covered with water, there is no problem. But if it's exposed to oxygen, the remaining water will turn to sulphuric acid. Already there have been reports that remote reaches of the lake have turned to battery acid. Clem Mason is not wrong: the lakes are facing a complete ecological breakdown even as its farmers contemplate ruin.

Clem tells me the story of Jervois, a small town up the river, sited not far from where it runs into the lakes. In the 1930s, riverside swamps were drained, levees erected, and the land turned into lush dairy farms flood-irrigated from the river. But now, without enough water to irrigate, the land has dried and cracked and the water table has fallen. If water is put onto the land, it is swallowed by the cracks and runs down into the increasingly saline aquifer. The land itself is warping and tilting, twisting into strange shapes as it dries. And as the river has fallen, the levees have also been drying out, cracks forming and growing large, the clay losing its ability to absorb water. A rush of water down the river could see the banks fail and the countryside flooded. Just a couple of years ago, sixty dairy farms operated at Jervois. Now there are four left. One is run by Clem's brother, surviving only because of the feed trucked in from Masondrina.

Clem returns to his work and I stroll past his landlocked jetty, the ground strewn with the shells of freshwater mussels, down to the side of the lake. The beach is of the finest white sand, stretching off to my left in a flat swathe hundreds of metres wide. A few cows amble down to join me, picking their way out to the water to drink. Backlit by the morning sun, it makes for a bucolic scene, the water so shallow it looks as if the cattle are walking on top of it. I photograph them for the record. For this scene cannot last. The whole lower lakes system, its environment, its agriculture, its engineering, is collapsing. Change is coming. The only question is: what sort of change?

From Masondrina I continue along the Narrung Peninsula. I'm on the eastern side of the lakes, the less populated side, the furthest from

Adelaide. The peninsula juts westward into the water, a solid triangle, its northern shoreline on Lake Alexandrina and its southern shore on Lake Albert. At the tip of the peninsula is the Narrows, where water flows between the two lakes. I drive along a shoreline of mudflats, impossibly wide beaches and sand islands. It's as if the tide has gone out, a king tide, and I need to constantly remind myself that there hasn't been a tide in the lakes for seventy years. At Point Malcolm I walk to a lighthouse—the only inland lighthouse in Australia. It operated between 1878 and 1931, guiding the paddle steamers that plied the lakes, but now it's long abandoned. Nearby, the light-keeper's cottage stands empty, door open to the elements. The roof is full of cooing, shitting pigeons. Outside, from the high ground beside the lighthouse, I can see more rain heading across the shrinking lake. Can there be anything as sad and solitary as a deserted lighthouse?

Just below it lie the Narrows, the link between the two lakes. The Narrows are closed. A crude earthen weir, a dyke, has been bulldozed into place, and on top of it sit seven huge pumps, their black pipes reaching down into a narrow channel dredged into the flank of Lake Alexandrina. The pumps are working, throwing out a low roar, as they transfuse scarce water from Alexandrina to keep Albert alive. The life-support costs millions of dollars each year. According to local lore, no doubt false, the pumps come from Cubbie Station—the government got them for a song because they're no longer big enough to meet the voracity of Cubbie's vampires. A ferry traverses the dredged channel. As I cross, I look beyond the makeshift weir and the pumps, trying to catch a glimpse of Lake Albert. But I can't see it. Once, the rushing water would have kept this strait free of silt, pushed by the shifting winds and, before the barrages, the tides. Paddle steamers and fishing boats came and went for more than 100 years. Now the pumps are pushing out silt as well as water, and sand islands, covered in bulrushes, have emerged to clog the old channel.

The rain hits with renewed fury. I wind my way carefully round to Meningie on the eastern shore of Lake Albert, not far inland from the Coorong. It's a quiet town: quiet now the holidays have passed, now the water has withdrawn. I shelter in a cafe where a chalkboard menu reveals the house specialty is mullet, freshly caught in the

Coorong. I put thoughts of Renmark behind me and brave it. It's delicious. Behind the counter is a large lime-green triangle. A quarter of a century ago I had a similar triangle stuck on the back of my car—back in 1983, the words inside it read 'No Dams'. The design has been appropriated by every environmental cause since. This one reads 'Save the Murray'. I can't disagree with the sentiment, but how to achieve it? The 'No Dams' campaign had a single, simple objective: to stop the construction of the Gordon-below-Franklin dam in the wilderness of south-western Tasmania. The protests stopped the dam, projected Bob Brown onto the national consciousness, and helped get Bob Hawke elected. The sight of the lime-green triangle, recycled for its umpteenth campaign, evokes a sense of nostalgia. Not that I was deeply involved in the Franklin campaign; the sticker on my car was more a statement of fashion than of deep-seated conviction. Rather, I'm nostalgic for its simplicity: stop the destruction of wilderness; no farmers displaced, no communities killed off, no real pain inflicted. It was a simple either/or equation—build the dam or don't build it—not the complex interplay of the Murray-Darling, not the global intricacies of climate change. 'Save the Murray'. Absolutely. But unlike the Franklin, the remedy is unlikely to fit within three black lines, no matter how big the triangle.

The bloke behind the counter is in a chirpy mood. Indeed, he can't stop chirping. It's the rain. It's the first of the year, a year already one-third spent. He can't stop saying how marvellous it is, greeting every squall that hits his plate-glass windows with another chirp. And he's convinced this is the turning point, that the rain will keep coming, that the cycle is changing. 'The pelicans are nesting in trees', he tells me, ticking off omens on his fingers. 'The Christmas beetles have come back. They've been around for a few weeks. Ants are moving to higher ground. And other water birds, ones that haven't been here for three years, have started returning.' I peer out the window as another squall washes up the street like a wave. It's certainly bleak enough. Maybe he's right. Maybe the weather is turning for the better. Then I regard the green triangle once more. If only it were all that simple.

A simple solution has been proffered to save the lower lakes: open the barrages and let in the sea water. The logic is plain. And seductive.

Before the barrages, the lakes were open to the Coorong and the sea. When the river was in flood, any salt water in the lakes would be flushed out into the Coorong, turning the lagoon into a mixture of fresh and salt water. Water would rush in and out of the Murray's mouth, depending on the river flow and the tides, keeping the mouth open and preventing it from silting up. If the lakes were low, sea water would flow in and keep them topped up until another flood came to push the salt back out. So the proposal is to return the lakes to their natural state. Simple.

Garry Hera-Singh doesn't think it's that simple. He's been fishing the Coorong and the lower lakes all his life. Not today though: the weather is too bad, the sea too rough, the winds too unpredictable. So while the rain tumbles outside, we sit in the dining room of his modern Meningie home drinking tea. He recounts a life shaped by the ebb and flow of the Coorong and the rise and fall of the lakes: how both his grandfathers were fishermen, how he loved going out with them as a kid and worked for one of them in his teens, how he bought his own licence in his twenties. That was in 1984. Since then he's witnessed the death of the southern Coorong, kilometre after kilometre lain to waste, mutated into a hyper-saline marine desert. Year by year, as less fresh water has crossed the barrages and the small freshwater channels that once drained neighbouring farmlands have been dammed, the Coorong's creeping death has edged further north.

'There's been no commercial fishing in the south lagoon for twenty years', Garry says. 'The only thing that lives in it are bloody brine shrimp. It's so salty now that even the aquatic weed has died. Fishing has been concertinaed into a third of its original areas. We fish roughly half the north lagoon. And the only reason we're still there is the dredging program that has maintained a marine environment. The dredging program is life-support for the Coorong.'

Lake Albert has also been lost to the fisherman, at least temporarily. Despite the emergency pumping from Lake Alexandrina, the water has fallen so low that gaining access is too difficult, as is navigating it without getting wedged on a sandbar. Fishing in what remains of the

Coorong and Lake Alexandrina remains good for the moment, but a sharp decline is expected in coming years. Fish stocks burgeon when the river is flowing, as nutrients are washed down from the Murray. Stocks peak three or four years later as the fish reach maturity. But it's not the short-term cycles that worry Garry; unlike a farmer, at least he gets advanced warning. It's the long-term that concerns him. He says there hasn't been a really significant flow of fresh water into the lakes and over the barrages since 1996. 'The habitable area has diminished extensively for the fish and the birds. And with the severity of this drought, that's been extended. The amount of birds that used to inhabit the Coorong, it's not even a tenth of what it was historically. And so where there were 80 000 to 250 000 birds per annum, you're now looking at between 5000 and 10 000.'

It's not all bad news. A new market has recently opened up for the once-derided Coorong mullet. Trendy Sydney restaurants have started putting it on their menus and are willing to pay three times the going rate. It brings to mind the fashion a few years back for orange roughy and Patagonian toothfish: the gastronomy of the endangered. Does scarcity itself hold an appeal for the inner-city gourmands, their palates bored with abundance? The Coorong mullet are not threatened, not yet, but perhaps reports of the Coorong's imminent demise have wetted inner-city appetites. Another cheery thought.

At fifty-two, Garry may not know much of the swirling currents of urban fashion, but he knows the waters of the lakes with an intimacy bred of decades of daily contact. His face is weather-reddened under his blue cap and his stubble is grey. During the warmer months he's on the water by five in the morning, taking fish from his nets. Then it's back to shore with the catch and to clean the weeds from his nets. By the afternoon he's back on the water. The water in the Coorong is so warm—up to 28 degrees—that he's constantly clearing the nets. 'It's a large shallow basin, almost tropical. Any longer than four hours and the fish are buggered', he says. Yet every day is different, his destination determined by the weather, the movement of the fish and the subtle interplay of a hundred other factors, both ordinary and arcane, self-evident and subliminal. We can all examine maps and read reports, but Garry Hera-Singh knows the waterways in ways more profound and more personal. It doesn't mean he's right, but it means he's worth listening to.

I suggest that opening the lakes to the sea would benefit fishermen: better a living marine environment than a dead freshwater one. It would prevent acidification, and the tidal rush between the Coorong and the lakes would keep the Murray's mouth and the Narrows open. But Garry disagrees, explaining the problems various seawater proposals would create. The first proposal is a holding operation: let just enough seawater into the lakes to cover the acid sulphate soils. Then, when the drought breaks and high river flows resume, the salt water would be flushed out again. But Garry says that if high flows don't resume, then the salt water will evaporate and the lakes will become increasingly saline as more and more sea water is introduced to top them up.

The second, more dramatic option is to open the lakes fully to the sea, so the tides can wash in and out of the lakes. This would prevent acidification and the lakes would not become more saline than sea water. But this proposal would require a massive weir to be built across the Murray where it enters the lake, to preserve Adelaide's water supply. This is not idle speculation: the South Australian Government has already started preliminary work on the weir, which would cost more than $100 million to build. The weir would preserve a pool of fresh water in the river, and prevent salt water from the lakes pushing up into it. Look at a map, take a bird's-eye view, and it makes a certain amount of sense; not a perfect solution, but a practical one. It makes sense because, on a map, it's easy to see where the river ends and the lake begins. But Garry Hera-Singh doesn't view the world from altitude; he sees it from sea level. He floats on the surface and contemplates what occurs in the depths below. For Garry, the dividing line between the river and the lakes is far more fluid. The land is so flat and the water so shallow that the wind pushes massive amounts of water up and down the river, amplifying its flow and keeping it from silting up. According to official figures, when the water at the barrages is 75 centimetres above sea level, it raises the water level 270 kilometres away at Blanchetown by 50 centimetres. The Murray south of lock one is really a long, thin extension of the lower lakes. Garry says hundreds of thousands of tonnes of salt flow past Wellington each year. If the weir is built, what's going to shift all that salt?

Garry leaves me with one more argument. It's not just salt that gets flushed out of the river into the lakes. There are also nutrients. 'Lake Alexandrina has been a nutrient sink for the whole basin for 10000 or 20000 years', he says. 'If they left the barrages open and the lakes filled with sea water, that water will be clear. The sunlight will reach the lakebed, causing photosynthesis. The weeds will go ballistic. You'll have algal blooms and weeds like you've never seen before. Navigation would be all but impossible, not just for us and our nets, but for the recreational boats as well. The smell would be unbelievable. The impact on the birds and fish would be enormous.'

⌣⟩

I walk along a wind-whipped lane, considering the conundrum. Having refuted the simplicity of sea water with the complexities of reality, Garry could only offer a substitute simplicity: send more fresh water down the river. If only. The wind is gusting in off the Coorong, wet with sea spray and thick with salt. I've spent the morning driving along the shoreline, dodging rainsqualls, braving short walks between cloudbursts. First, I went down to the Needles, the thin passage of water only 100 metres wide connecting the north and south lagoons. The name derives either from the narrowness of the passage or the razor-sharp reefs that render its navigation precarious. Take your pick. I took a cautious sip of the water. What can I tell you? It was salty. From the shore, there was no discernible difference between the short frothing waves of the two lagoons. And yet there was a difference, even to my city-bred eyes. To the north, in the shallows, there were pelicans and gulls, but not to the south. Next, I ventured back north alongside the water, forced to stop once along the way by rain so heavy and horizontal that I couldn't see. I'd passed Garry's unattended fisherman's hut, and then a hapless young couple trying to erect a tent, as if practising origami in a wind tunnel. Finally, the public road ended near Pelican Point, blocked by a padlocked gate. I took a chance and left the car, hoping a break in the weather would hold. I'd pushed through the rusting pedestrian gate and proceeded on foot.

Now I'm trudging along a thin strip of bitumen surrounded by low tough gorse, hoping to reach the Tautwitchere barrage before another itinerant tempest soaks me through. The countryside here is low and windblown, with stunted trees permanently angled away from the prevailing weather. A couple of sparse pines accentuate the bleakness rather than relieve it. Across the choppy waters of the Coorong, foam-flecked and milky green, I can see the sand dunes separating the lagoon from the ocean. They're splotched with low green vegetation but there are no trees: too much sand, too much salt, too much wind. Somewhere ahead is the barrage, the longest of the five barriers separating the Coorong from the lakes.

As I walk I ponder the complexity of the lower lakes system. No part is truly separate from the other; they're all interconnected. Another metaphor. From the moment the water spills over the weir at Blanchetown, what affects one part of the system ultimately has an impact on the rest. And the amount of water flowing over the weir is determined by how water is used and abused throughout the entire Murray-Darling Basin. Water extracted for cotton in Queensland eventually has some impact on the lower lakes, no matter how miniscule. And ultimately, the amount of rain falling in the watershed of the Alps, the amount flowing into Hume and Dartmouth and collecting in thousands of other dams, weir ponds and farm storages, is being determined by shifting weather patterns and changing climate. The complexity is immense, unknowable. What meteorologist can predict the weather, even a month from now? What hydrologist can explain the mysteries of the aquifers and artesian basins, how and why and where they flow, how they interact with the rivers? What politician can decide which farmer to evict and which to support, which forest is valuable and which is not? The systems are cosmic in scale. An engineering solution imposed with the best intentions in one location can create unforeseen consequences elsewhere in the system. I wonder if anyone can properly get a perspective on the

Murray–Darling. I doubt it. Everywhere I've travelled along the river, I've encountered people confident in their own opinions: the farmers, the environmentalists, the front bar pundits, the locals always thinking someone somewhere else is taking their water. The technocrats in Canberra have perhaps the best oversight, but they have a bird's-eye view of the system, ignorant of the perspective from ground level, unaware of deeper currents. I've seen how inadequate and overconfident bureaucrats can be when it's not their livelihoods that are on the line, not the trees in their own backyards that are dying. Overlay that with the compromises and limitations of politics, another dimension of complexity; with senate trade-offs and interstate haggling and election timing. There are no simple answers.

After walking for twenty minutes or so, I reach the beginning of the barrage. It's a long elevated concrete roadway, stretching dead-straight before me and kilometres long, like a computer-generated illustration of perspective. The roadway is wide enough for a single vehicle, with metal handrails on either side. I walk a ways out, leaving the land and standing in the void above the division in the waters, like Moses with an engineering degree. To my right, Lake Alexandrina is all but empty, the reef rock exposed, ponds of still water gathered in the shallows, sheltered by the windbreak of the barrage. Sand islands straddle the middle distance, with a crane or two picking disconsolately at the sand, as if from habit rather than hunger. To my left, the waters of the Coorong heave and chop in frenetic contrast, a metre and a half higher than the lakebed. The short sharp waves attack the barrier: hungry, fierce and foaming at the mouth, as if sensing the vulnerability beyond, eager to claw their way through. Lake Alexandrina lies fragile, calm and unknowing. A famous quantum physics experiment comes to mind: photons passing through slits in a barrier, each inexplicably knowing what the others are doing. The interconnectiveness of nature. In the distance a huge darkness looms above the horizon. I turn and start the long walk back, already knowing the storm will catch me well before I reach the car.

The word 'patrician' keeps bumping around my head as Keith MacFarlane crushes my hand and invites me into the shearers' quarters. They're like no shearers' quarters I've ever seen, their hand-carved stone walls softened by rose bushes, the interior spacious and tastefully renovated. There is the smell of old money about Wellington Lodge, permeating even the outbuildings. Keith, now sixty-five, combines the no-nonsense practicality of the working farmer with the inherent confidence of the squattocracy. He and his wife have moved to the shearers' quarters, down past the full-brick shearing shed, so his son Richard and his young family can live in the grand old homestead. Keith's great-great-grandfather, Allan, founded the property in 1845. The family's wealth was built on wool: Wellington Lodge carried 45 000 sheep at the turn of the last century, now it carries none. Keith and Richard decided a few years back that there was no money left in wool, choosing to concentrate instead on Angus cattle and cropping. But the sense of tradition is evident; Richard is the sixth generation, his young son, Tom, the seventh.

The property lies on the shore of Lake Alexandrina, where it meets the Murray. The proposed weir would stretch out from Keith's property. According to the bird's-eye view, he would be one of the winners: Wellington Lodge would be able to pump from the fresh-water weir pond, its level guaranteed and protected from the newly saline lakes. But Keith MacFarlane is dead against it. He doesn't believe the weir will protect Wellington Lodge from salt. Just the opposite.

'If the weir does go in, the salt will build up', he says. 'It will back up the river and ruin our stock water. Five hundred tonnes of salt goes past here every day. The problem is that the river is so slow moving … If the salt can't get past the weir, and believe me it won't, it can only do one thing … It will start moving further and further up the river, right up to where they take Adelaide's water and beyond.'

Keith holds a licence to pump water from the lake, but he irrigated his last crop two years ago. His pumps no longer reach the water. He would need to dredge a new channel to access it. With his allocation cut to 18 per cent, the same 18 per cent Tim Whetstone has been granted up in Renmark, he doesn't believe it's worth it. Yet he steadfastly refuses to lease out this year's allocation. 'I've only ever leased it out on one occasion, when we weren't using our full entitlement. I said I would do it with one stipulation—that it didn't go out of South Australia. But it did. It ended up at Tandow Station [on the Darling]. Water should not be allowed to be traded out of the state.'

It's a principled stance, one that sits uneasily with technocratic models that assume self-interest and 'rational' economic choice. Just before I leave, Keith tells me about his family history, how they still have his great-great-grandfather's diaries. He says the National Library has expressed interest in them, but he'll be damned if he's handing them over to 'Canberra'. 'It's our history. Why the bloody hell should I? We've got a seventh generation coming through. They can sod off.' The word patrician remains resolutely ensconced in my mind as I drive back down the drive, past the Angus cattle grazing among the low gorse, back to the highway.

⌒

Langhorne Creek is pastoral-perfect, all sandstone cottages and autumn leaves, with a fire in the hearth and this year's vintage stewing in stainless-steel vats. The town is bed-and-breakfast heaven, surrounded by wineries ranging in scale from the boutique to the industrial. Laid out along the road are piles of large PVC piping, as if Santos has struck a new oilfield. But the pipes aren't for oil. The Bremen River remains steadfastly empty despite the rain, and although Langhorne Creek sits only a few kilometres from the western shores of Lake Alexandrina, the lake has receded so far that the vineyard pumps can no longer access it. The PVC is for a new pipeline that will carry water from more than 40 kilometres upriver. It will cost more than $100 million and be operational by next summer.

Throughout the basin, nearly every farmer I've met believes this is a drought that will end, as all droughts end. But at Langhorne Creek, the pipes tell a different story. The growers are turning their backs on the lakes; they don't believe they will recover. The 110-kilometre-long PVC web is being largely funded by government, but will be owned and operated by the irrigators themselves. Communalism is not dead in South Australia. But the government funding is instructive.

It's costing too much money for a temporary fix; the groundwork is being quietly laid to abandon the lakes.

I call into a small winery, parking near the elegant old stone homestead. The cellar door is up a wooden ramp, on the second storey of a more utilitarian building. The elevation provides a view out across the vines. But there is vacant land as well, ploughed to expose the rich dark soil of the flood plain. The woman behind the counter, one of the owners, explains that the vacant fields are for horseradish but that this year's crop has been reduced, to save what little water they've had—groundwater—for the vines. She tells me that connecting the property to the new pipeline will be expensive, costing around $45 000. But it will be better than this year, with their 18 per cent allocation sitting out of reach in the lake. 'We can sit on it, we can lease it, we can sell it. We just can't use it', she says.

I try some of the wines. They're very good, even the young ones: elegant, smooth and full of promise. I sample a durif: as big as Texas and just as bold, yet with a hint of sophistication. I buy a varied dozen and watch the moving sun and shade alternate across the somehow anglicised landscape. I ask if they've sold this year's water, the water they can't access, in order to recoup some of their costs. The woman tenses for a moment. 'Oh, no. We won't sell it. We're not going to give it to those bastards upstream.' And then the studied elegance returns.

⁓

At Milang, on the western shore of Lake Alexandrina, the sign on the jetty says 'No jumping, no diving, no discharge of offal, blood or offensive material into the sea'. Which is doubly peculiar, for this isn't the sea—it's a landlocked freshwater lake. But the error is under-standable enough. There is the unmistakeable feel of the maritime here, down to the barnacles clinging to the piles of the jetty. The old pier, its wooden planking uneven and unsure, stretches 100 metres or more across the sand to nowhere. At its terminus there is nothing but mudflats. The lake proper is at least half a kilometre away. Who knows how long it's been since the last boat pulled in here.

At Clayton, further round the shoreline, the jetty remains connected to the body of the lake, but only because of dredging. Otherwise it's sand and mud. The lake appears to funnel back into a river here, as a residual channel of the Murray curves around the western side of Hindmarsh Island, heading towards Goolwa and the barrage that prevents it reaching the Coorong and the Murray's mouth. It's not the only place where Lake Alexandrina is prevented from reaching the Coorong: other barrages reach out from the eastern side of Hindmarsh Island. Indeed, the island is the cornerstone of the barrage system. Looking at the island across the channel from Clayton, I can see a substantial bungalow hugging its shore. I look more intently: it's not a house but a houseboat, marooned on the mudflats.

I reach Goolwa, the port developed all those years ago by Francis Cadell and the founders of South Australia. Goolwa's industrial days are long gone, but it remains the undeclared capital of the lower lakes. An hour's drive from Adelaide, it has become a beachside town, a tourism town, a retirement town. It's a comfortable place, with three bakeries, two pizza joints and a DVD store. The main street is lined with trees and heritage buildings of stone and wood. Down by the port, there's a brewery, a cafe and the obligatory paddle steamer. There's a railway station, where a steam train terminates every weekend after an 11-kilometre trip from the sea at Victor Harbour. More rain comes billowing in from the south-west and I shelter in the cafe. I look out at the Hindmarsh Island Bridge, a massive span of concrete towering above the harbour. The bridge, elegant enough in its arc, looks ungainly and overly tall, with its hydrodynamic footings stranded above the shallow water, revealing the smaller piles below. It looks like a car that's been jacked up and its wheels stolen.

There are signs everywhere in Goolwa, not the lime-green triangles of Meningie, but scarlet rectangles brandishing white letters that declare 'Goolwa needs water now!' The difference goes beyond the colour scheme. At the entry to Meningie, a town of locals, a large banner has been erected declaring 'NO DAMN!! WEIR!' But at Goolwa, a leisure town full of Adelaide weekenders, the demand is less specific. The town simply wants water, and isn't so fussy whether it comes down the river or across from the sea. Along Riverside Drive, 'for sale' signs are dotted out front of the luxury holiday homes of Adelaide's well-to-do. The road winds along to the Goolwa barrage that stretches across the channel between Hindmarsh Island and the mainland. Not far from the barrage, a jet-ski-hire place has been left high and dry. Someone has painted 'LET THE SEA WATER BACK NOW' in angry blue letters across a sign saying 'group bookings welcome'. A smaller sign says 'Closed until water levels rise by 2 metres'. On the shopfront itself there is a sadder message: '4 Sale. Walk In Walk Out. $100K. Jetski Hire. Café shop. 5 Yamaha jetskis. On-site van. All wetsuits, life jackets. Quad bike. Tools etc'. It doesn't seem a lot of money.

I drive back past the port to Liverpool Street, a waterside road running to the north, lined by more-modest holiday homes. I stop at a small marina and walk out onto the crisscross of wharfing; the piles are covered in barnacles, bleached white and dead. Below me are the boats, two dozen of them, lying in the mud, as if waiting for the tide to return. There's *Hoodlum* and *Dolphin* and *River Nomad*; *Sunrise* and *Morning Mist* and *Grey Dawn*. Every day they sink a little lower in the mud. Towards the end of Liverpool Street, a few blocks inland, I find a motel: the Riverport. Out back of the rooms there's a swathe of lawn, equipped with barbecues. But it's marred by a large and muddy hole, lined by treated pine decking, with a couple of ducks paddling about in the puddle in the bottom. At first, in the dim late-afternoon light, I mistake it for some half-finished swimming pool, the lining put on hold because of some unexpected subsidence or until business picks up. 'Overly ambitious', I conclude. It must be two-thirds the size of an Olympic pool and twice as deep. Only when I approach the edge do I see the jetty, that the hole is open at one end. It's a dock. Disoriented by its location, I'd thought the name Riverport was merely figurative, like the Bridgeport Hotel in Murray Bridge. Curious, I follow the narrow channel leading off from the hole. It's completely empty. Fifty metres along, it joins a larger channel, a little stagnant water in its bottom, which leads me to the left. On the far bank are houses, rather grand houses. There is a prominent yellow home, bordered on one side by the main channel and on another by a branch. It's a canal development: 'Riverside Cove. Authorised Access Only'; 'Private Waterway. No wake'. No wake is right; the canal is nothing but a muddy ditch, the private wharfs hanging out over nothing but air. The houses look locked up and desolate. I push on, following the main channel. There is a little water in it now, but it's nothing but a remnant topped up by the rain. The canal turns to sand and I jump down and walk the rest of the way along its empty bottom. It breaks out into where the lake once was, now a sandy beach. A hundred metres from the waterline, a pole is planted in the sand. 'Speed limit 4 knots' reads the sign on the top. There's a depth gauge running down the side. I look up at the 2-metre mark. Further along I find a concrete block, galvanised chain still connected, all that remains of a mooring. I feel like an archaeologist, combing through the remnants of some ancient disaster, the detritus of a lost civilisation.

Another day, and still the rain sweeps in from the Southern Ocean. I escape to Adelaide and visit an old friend from Bathurst, Gemma, and her husband, Richard. We eat a home-cooked dinner, drink too much wine, and dissect the strange rituals of the locals. Gemma, from Sydney, and Richard, from Canberra, say it took them six months to get used to drinking the water. Now they drink it straight from the tap, unlike many of their friends, who use household filtering systems. There are plans to build a hugely expensive and energy-intensive desalination plant. Adelaide isn't alone. Sydney, Melbourne, Brisbane and Perth all have similar ambitions. The only capitals not running short of water are Darwin and Hobart. One Adelaide council area, Salisbury, has set an example by recycling storm water, using small wetlands as filters, storing water in aquifers, and even recycling a limited amount of sewage. This last is for irrigating playing fields and the like, but not for drinking. Drinking recycled sewage is taboo. When the Toowoomba City Council in Queensland proposed introducing it to the household supply in 2006, more than 60 per cent of residents voted it down at a referendum, a lesson to politicians everywhere. After dinner at Gemma and Richard's, back at my motel, I guzzle tap water, belatedly attempting to dilute the wine and avoid a hangover. The water tastes better than at Murray Bridge, but nowhere near as good as the recycled sewage I've drunk in Singapore. Where do the people of Adelaide think their Murray water has come from? There are at least two million people living upstream of Adelaide. I know where Canberra's effluent goes, well treated though it may be: straight into the Murrumbidgee. Perhaps that's the strange taint I can taste in the water: familiarity.

⌒

The weather at Goolwa is still lousy. I call in at the tourist information centre, perched on the hill overlooking the port, and ask about getting down to the mouth of the Murray. After not quite reaching the source with Cameron, maybe I can make amends by getting to the river's end. But I'm told I would need to undertake an eight-hour return hike along the sand hills separating the northern Coorong from the ocean. A fresh cloudburst buckets into the side of the building, rattling the windows. An eight-hour walk. Terrific. The woman says if I don't want to walk, I can get a clear view of the mouth from Hindmarsh Island. I decide discretion is the better part of valour and head over the bridge, hoisted on its spindly legs above the residual lake. After the bridge, the road advances straight ahead, past the entry to a marina development and up the spine of the island. At the top of the rise, a monument bears a verdigris-powdered plaque erected in 1930, commemorating the centenary of Charles Sturt's expedition. This was the end point of his exploration. From the heights of the island he could see across its flatlands, across the end of the lakes and the beginning of the Coorong, across the sand hills to the ocean. It was here he finally accepted there was no inland sea. The rain has stopped and a burst of sunlight breaks through. I regard much the same view as Sturt: there are no buildings to dispel the illusion, just a road and a couple of fence lines. Then I go further than Sturt, driving down to where I can see the mouth, although not as far as Sturt's friend and colleague, Collet Barker, who reached the mouth itself a year after Sturt, and was then speared to death for his efforts by an irate local. Breakers are crashing against the sandbar at the Murray's end. It's only a kilometre away across the Coorong: shallow, shoaly and shifting blue in the sunlight. The ever-present dredge is off to one side. I take a couple of photographs, but in the end there's not a lot to see. It's a little anticlimactic.

Heading back towards Goolwa, I turn into the island's marina development. Its scale takes me by surprise. It's huge. I learn there are 16 hectares of basins and lagoons alone, with hundreds of luxury waterside homes already built and more on the way. They stretch out along the canals that spider out from the harbour, each one with its own berth. It's the largest freshwater marina in the Southern Hemisphere. Now I understand what the controversy over the Hindmarsh Bridge was really all about: real estate. How Australian. There's still some water lingering in the basin; all the waterways have been excavated to 3 metres below the normal surface level, leaving them about one-third full. Most of the shallower berths in the central marina have been evacuated, but there are still plenty of vessels to be seen. Most are motor cruisers, with a token sailboat now and then. It's that sort of place. Down by the harbour there's a restaurant and tavern, but I drop into the marina centre instead. The friendly young woman at reception tells me that, despite the water crisis, houses are still selling reasonably well. She says interstate buyers have dried up, those looking for a quick capital gain, but that they've been replaced by bargain hunters from Adelaide. They're looking for a holiday home, or something they can rent out before retiring here in ten or fifteen years time. With the city just an hour's drive away, I can see the appeal. A retirement development is underway; so is a gated community for the over-55s. I ask her what should be done about the receding water. She says all the residents want water, but most don't care whether it's salt water or fresh water. Many believe salt water would be an improvement.

I drive round the streets, looking at all the brand-new houses. To call them McMansions would be unfair. For the most part they're well spaced, set back from the canals. But there is a certain sterility, an antiseptic aesthetic designed not to offend. The international design school of beige. But to each his own. I guess it will be more inviting when the trees grow and the boats return. For there are no boats along the canals. Not one. Every house has a jetty, its own private berth standing high and dry above the shallow water, but all the boats have been removed. It gives the development a bare, half-finished look. At least the water is still there, no matter how shallow. If the water went altogether, and the residents were looking out across muddy drains, any residual appeal would vanish very quickly.

I pull over at the top of Excelsior Parade, near one extremity of the subdivision. More rain is blowing in, so before it hits I rush to photograph one of the sterile canals. And I finally achieve what I've been threatening to do this entire trip: lock my keys in the car. The rain starts spattering. Oh joy. I do up my raincoat and trudge along to the nearest house, praying someone is home. A teenager opens the door. I ask him for a coat hanger; the car door isn't completely closed and I'm hoping I might be able to trip the lock. But I've lost the knack and ten minutes later I'm back to find the road service number. This time a slim bottle-blonde woman in her early forties answers the door. She invites me in, finds the number for me, insists I use her phone. The RAA reckons they'll arrive in an hour. Maybe two. More joy.

The woman's name is Bronnie. She has a husky voice and a no-bullshit accent. She's sitting at the dining-room table with a pile of old clothes and a laptop. Until February she was working as a caterer up at the Prospect Hill tin mine in the Flinders Ranges. It was fly in, fly out: two weeks at the mine and a week off in-between. She would drive from the marina to Adelaide airport, leave her car there, and pick it up a fortnight later. But the economic crisis and plummeting commodity prices closed the mine. She doesn't miss the work, but she misses the money. Now she scours op shops for second-hand clothes, reselling them on eBay.

Bronnie's house is sited on a rise at the end of a canal. Through the picture windows, we can look down the length of the canal, with its impressive houses and empty jetties. She tells me she and her partner, Adam, owned a boat but it ran aground, the rocks ripping into its hull. So they sold it. There hasn't been a boat in their canal for a couple of years. Bronnie says Adam and a few of his mates are thinking of buying remote-controlled model boats. They plan to sit out on the jetty with a couple of beers and race the boats up and down the canal. She says the wives and girlfriends are thinking of setting up a hot tub so they can have a few drinks and watch.

Adam wanders in. He's a big bloke, a plumber, with a mullet and a lazy eye. I ask about all the 'for sale' signs that are dotted about the marina. The couple aren't bothered, saying most are new places constructed by spec builders, being put on the market for the first time. But they point out a substantial house a few blocks along the canal. 'They had it on the market for $600 000', says Adam. 'They were offered $500 000 but knocked it back. Now they've dropped the asking price to $535 000. The way things are going, they may have been better taking the 500 grand.' I ask the couple whether they think that sea water is the answer to the marina's problems. Bronnie is against it, but Adam is more equivocal: 'Two separate real-estate agents have told me that if they let the sea water in, property values would skyrocket.'

The teenage boy returns, a friend of Adam and Bronnie's own son. He's straightened out the coathanger and wants to have another crack at the car door. The rain has abated, and within about five minutes we've opened it. I return to the house, cancel the RAA call-out, and thank Bronnie and Adam for their hospitality. I head back out of the marina, back over the bridge to Goolwa. And as I drive, I consider all the money tied up in the development: hundreds of millions of dollars. I pass a couple of the red and white signs: 'Goolwa needs water <u>now</u>!' It will get it too: the weight of so much money is unlikely to be repudiated. It may be sea water, or it may be fresh water. There's a new proposal to build a network of levees that would extend the main course of the Murray down the western shore of Lake Alexandrina, joining it to the channel that separates Goolwa from Hindmarsh Island and leads past the marina to the Goolwa barrages. This channel already carries 70 per cent of the water that reaches the Coorong in good times. Goolwa and the marina millionaires would have their water, but the lower lakes would be cut adrift. The people of Meningie would be left to their fate, but I can't see the old farming and fishing communities of the eastern shore competing with the big-city money of the western shore. How can a family-owned caravan park in Meningie compete with the Hindmarsh Island marina? I've got a feeling the Murray-Darling is about to get another engineering solution.

There are signs the weather is starting to break. The frontal bands sweeping in from the south-west are becoming more widely spaced, losing their intensity. The sun is starting to reassert itself. I drive down Goolwa's Riverside Drive, past the holiday homes of concrete and plate glass, past all the signs, down past the abandoned jet-ski-hire place, down to the barrage. I park the car, taking care to remove the keys. But instead of walking to the barrage, I turn to a duckboard path heading south across the dunes. I cross one last road then walk into the dunes proper. The path climbs up over a low ridge and down into a sandy gully, where it carves through the middle of an Aboriginal midden. It mounts another rise and I can see the ocean, smell the sea mist and hear the crashing waves. A few more minutes and I clamber down onto the beach itself. Journey's end. Not the Murray's mouth but, like Cowombat Flat, the best I can do. The beach is narrow, pounded by a ferocious surf, the sea more white than green. The thin intermittent band of sand stretches in a long straight line in either direction. There are neither headlands nor inlets here: just the ocean running hard against the dunes. There's not another person in sight, just a couple of gulls, hovering in a glide above the beach, held aloft by the south-westerly.

I consider my days along the river, what I've learnt. The obvious and the elusive. There's not enough water to go around, and there's not likely to be enough anytime soon. Even if the drought breaks, even if there were to be a major flood, the old belief that Hume and Dartmouth and the Snowy Mountains Scheme have drought-proofed the Southern Basin has gone for all time. Even if the dams were to be miraculously filled, the confidence to empty them again has been destroyed. And that's the best-case scenario: the dams refilling and the water being used judiciously. The worst scenario is that the drought continues, revealing itself to be something entirely more sinister: the first wave of climate change. If so, the weather of the past few years would become the norm, and what used to be the norm would become the exception. We need to decide what can be saved. And there is so much worth saving: the vestiges of the great river red gum forests; the pioneering communities that inspired the ethos of the bush and our national identity; the inland sea, perhaps our greatest civil and social engineering achievement. I look at the implacable surf and wonder whether our political system, with compromise being both its greatest strength and its greatest weakness, can deal with such a challenge in time.

The federal government's $12.9 billion water plan is not without merit. It will eventually set extraction limits on river water and groundwater. It will establish environmental flows. Given the constitutional limitations, perhaps it's as much as we dare expect. But it's not enough. And it's not quick enough. If the lakes are to be saved, then they will be saved by rainfall, not by governments. I guess I could have come to similar conclusions sitting in Canberra: reading reports, weighing statistics, consulting experts. The bird's-eye view. But I'm glad I haven't. There are some things that should be learnt at ground level—walking the empty riverbeds, experiencing the dying wetlands, meeting bewildered farmers. And drinking wine on a fine summer evening, and walking through the high clean air of the Alps, and seeing the wide-eyed joy of children watching fireworks at a country show.

Australia is changing. The great inland basin of the Murray-Darling system will never be the same again. The great utopian age has come to an end, the age that began with the aspirations of the Chaffeys in the 1890s, stretched through the construction of the inland sea between the world wars, and culminated in the nation-building ambitions of the Snowy Mountains Scheme in my parents' generation. The great push into the inland has reached its high-water mark and is beginning to recede. There will be no new Snowy Mountain Schemes, no new irrigation areas; there is not enough water. The people are retreating to the garrison towns, buttressing them against the unpredictable challenges of climate change, market forces and political expediency. The countryside my father knew is coming to an end, beginning to fade like a Mallee town or an ephemeral lake. I wonder what countryside my children, Cameron and Elena, will know. The river communities will endure: the economic imperative is too great, the resilience of the people too formidable, the needs of the environment more and more appreciated. I hope the communities retain some of their vitality and their authenticity, that not too much fades or is lost to heritage-listed theme parks. I hope the people of the bush will continue to spread out along the rivers on their holidays the way the rest off Australia spreads along the coast.

I'm still not sure what I've learnt from my journey, other than I'm so very glad that I undertook it. In years to come, I'll be able to recount what life was like on the river that summer when I took the time to travel it, to tell of the things I saw and the people I met. I will be able to say I saw it at its peak, or near enough to it, and could appreciate what was achieved during the great utopian age. I'll be able to tell stories of the environment, so resilient yet so fragile; of the farmers, so practical yet so delusional; of the towns, so friendly yet so hidebound. I'll recall the towns with their vision statements: 'Cowra: World Centre of Friendship'; 'Conargo: the Legend Lives On'; 'Cullulleraine: Enjoy the Magic'. And their Chinese restaurants: the Rising Sun in Dubbo, the Imperial Chopsticks in Albury, and Rothman's All Joy in Wodonga. I will boast that I saw the Cunnamulla Fella, the Dog on the Tuckerbox, Rameses II and a white elephant trumpeting in the rain on a rooftop in Murray Bridge. I'll boast I saw it in its glory, in its pettiness, and in its doomed aspirations. But now time moves on and change comes, discreet and thunderous. I wonder what this new age might be called.

The sun is moving behind another band of cloud. Another squall is building out at sea, grey and ominous, a final wave of rain bearing down upon me. It's time to go. I take a last look at the ocean and begin my retreat, heading back into the hinterland on the long drive home.

Epilogue

'Speed Limit 4 Knots': the main channel at Goolwa on the lower lakes

It's November, a year to the day since I started off into the Murray-Darling. The spring thunderstorms have come marching through Canberra, spurring an explosion of growth. Walk by the city's lakes or through its bushland or around its suburbs and all is green, as if the drought never existed. I've already had the Rover out twice, and

I may have to give the lawn a third mauling before summer arrives to kill it off. Yet the spring green has also revealed those plants that have failed to survive last summer's heat. The silver birch that dropped its leaves during February's heatwave is dead. The verdure is illusory: the city's rainfall so far this year has totalled 288 millimetres, well below the 445-millimetre average. As well as tapping the Murrumbidgee for water, plans are proceeding to spend $363 million to expand a dam on the Cotter River. Stage 3 water restrictions remain in place: no sprinklers, no watering of lawns under any circumstances. The weather has been erratic: the warmest August night on record preceded the coldest October day for ten years. In-between, the city experienced the twelfth September in a row with above-average temperatures. Now the heat is on its way: the forecast for this week is for temperatures in the mid-thirties. The city's green won't last much longer.

It's been an interesting spring. In late September Sydney was cloaked in an eerie fog of orange dust blown in from the central deserts, a phenomenon not experienced for generations. With summer still weeks away, bushfires threatened homes across four states and, even as Victoria's royal commission continues to sift through last summer's disaster, authorities warned that this year could be even worse. A new six-tier fire-warning system has been introduced, so that sitting above 'low-moderate', 'high', 'very high' and 'severe' will be 'extreme' and 'catastrophic'. What's next: apocalyptic? In parliament, the debate over an emissions trading scheme drags on and on.

My manuscript is down in Melbourne, being copyedited, proof-read and typeset, polished and adorned for its public debut. But before I let it go altogether, I feel I should get on the phone and learn what has happened out along the river over winter.

Up on the Culgoa, Hugh and Pam Fennel have at last found a buyer for Boneda. After 126 years and four generations, they have just a few months more on the property. Their daughters want to come home one last time, for a final family Christmas in the bush. The new owner lives at Swan Hill, so the homestead will soon lie empty, like so many other houses in the district. There's some water in the river after 5 centimetres of localised rain, but almost none where the Thompsons are a few kilometres upstream. Ranger Bart Schiebaan tells me that rain earlier in the year gave the Culgoa National Park a temporary lift, but there's no more water in the river than there was last year.

At St George, Chad Prescott's storages are all but empty. He'll grow only the smallest of cotton crops this summer, using the last of his water. All of the area's flood harvesters are in the same position, and even the river irrigators are struggling. After only two decent summer floods in a decade, the water harvesters have hit hard times. Last week, the massive Cubbie Station went into voluntary administration with debts of more than $300 million. Chad says the district will feel the consequences: Cubbie is part of its economic base, a big spender and a big employer.

Out on the Paroo, the wild river's rude health continues to mock the eastern engineers. Jake Berghofer is still irrigating, and the New South Wales graziers are still trying to stop him. Ev Bartlett says the river at Toonborough has been dropping recently, but there's more water on its way. Fresh rains in Queensland will give it a good kick along when the water gets down to the property in about a month's time.

At Bourke, Steve Buster is still on Darling Farms. The sale he anticipated fell through and the family is still managing the property for the bank. But he's looking at the second good year in a row. Following on from last summer's cotton crop, a February flood refilled the property's huge on-farm storages, enough to water a winter wheat crop and this summer's cotton. The family has reopened its packing shed and restarted its cotton gin, and has also begun growing watermelons, providing much needed jobs for fading Bourke. The plan is to bolster the business and get the family a better return when the bank sells, although there's an outside possibility, just a glimmer, that a few more good years could put the Busters in a position to refloat the business. Yet the Darling River had stopped flowing by the end of September, and town water restrictions have been imposed for the first time in two and a half years.

Downstream at Menindee, the February flood filled Lake Wetherell, the smallest of the lakes, but no water was put into Lake Parmamaroo. It was half-full when I visited last December—now it's empty, a moonscape like Menindee and Cawndilla. Mike Arandt has retired and the operation of the lakes is now someone else's responsibility. There still hasn't been any decision made on which of the six water-saving options will be built.

Dartmouth, the Murray system's dam of last resort, has faired marginally better than it has for four years. But all that means is that it's 30 per cent full, compared to the historic average of 85 per cent, and Peter Liepkalns expects that by May it will be back down to 22 per cent, about the same as this year. He says the catchment is still like blotting paper; the system is still running on empty. At a meeting earlier this week, forecasters warned that south-eastern Australia could be heading back into an El Niño pattern and a repeat of the worst days of 2006 and 2007, except that this time it would be heading into drought without a buffer in the dams.

Out along the Murray, this year is shaping as a little better than last year for irrigators, but not by much. Mario at Orange World will get 97 per cent of his high-security allocation, but at Wakool the general allocation is just 10 per cent, compared to last year's 9 per cent. But all that government infrastructure money is helping: enough water has been saved by the irrigation company to give farmers an extra 4 per cent allocation. This has been enough for John Licato to sow a rice crop for the first time in four years, though it's just 20 hectares. It's not economic to produce such a small amount, but John is tired of doing nothing and wants to grow something. Even with such a tiny crop, he will need the allocation to increase as the season progresses; if it doesn't, he'll have to buy water.

Merv Membrey says Wakool is holding its own for the moment. Negotiations continue between the government and irrigators about shutting down 74 kilometres of canals, but so far none have been permanently closed. The landholders want out, but the government won't pay the asking price. The delay hasn't helped the school: the principal has been told she is too qualified for such a small place and will have to leave.

Time has run out for Andrew and Gail Tully. This year's allocation was the death bell. They have sold the last of their cows and given up any hope of dairy farming. A punt on a 160-hectare winter cereal crop hasn't paid off: lack of rain ensured the harvest won't cover costs. The couple sold this year's water allocation as soon as they received it.

They're now contemplating selling the farm altogether, along with its state-of-the-art dairy.

The Barmah Forest is turning into a national park—the legislation is before parliament. Logging finished for good in July. Kevin Swan says he will shut down his Picola sawmill next week, once he has milled the last of his logs. He and his son, Trent, have found work at Cohuna, an hour and a half to the west, collecting green firewood in the Koondrook red gum forest for another miller. There should also be some work available in the Barmah Forest assisting ecological thinning. Kevin hopes that between the two locations, he and Trent may yet be able to earn a living from harvesting wood.

Despite its imminent national park status, Sharon Terry says the forest is looking no better than when I visited. There's been minimal rain over winter, and no water has been allowed past the regulators into the creeks and forest proper. But she says the woodland is slowly recovering after the end of grazing.

In the South Australian Riverland, the hot weather is setting in. The forecast for next week at Renmark, three weeks out from summer, is 43 degrees. Tim Whetstone says the advent of hot weather is searing away the camouflage of winter, revealing which farms haven't survived as trees and vines denied irrigation brown and die off. Nevertheless, Tim says there is a degree of optimism among surviving farmers— their high-security allocations are up from 18 per cent last year to 46 per cent this year, with the possibility of another increase before the summer is out. It's not the 100 per cent they're used to, but many see it as a sign of better times ahead. Even so, Tim says the allocation is not enough to keep all the area's citrus trees alive. Sure enough, the day after I speak to Tim, it's announced that the iconic juice factory at Berri, 20 kilometres from Renmark, will close with the loss of sixty-four jobs. The company says the Riverland operation is no longer viable. Grapevines will cope with reduced water, but not with reduced prices. My local bottleshop is selling wine from south-eastern Australia for $1.99 per bottle. Who can make money from that?

I call Clem Mason at his dairy farm beside the lower lakes. He has what sounds like good news: the weir planned to block the Murray where it enters the lakes is on hold for at least another summer. But

Clem says that convinces no-one, that a decision was always going to be delayed until after the state election in March. Meantime, the pumps have stopped moving water from Lake Alexandrina to Lake Albert. Clem says the evaporating water in the smaller lake is becoming increasingly saline, and is now about half as salty as sea water. The pumps themselves have gone to protect Goolwa. A bank has been built right across the channel of the Murray between Clayton and Hindmarsh Island, a kind of reverse weir, forming a lake within a lake stretching from Clayton down to the Goolwa barrage. Water has been pumped into this inner lake from Lake Alexandrina proper and has flowed down Currency Creek and the Finniss River thanks to rainfall in the Adelaide Hills. The water level isn't as high as in the Coorong, but is noticeably higher than in Lake Alexandrina and along the Murray as far 'upstream' as Blanchetown. The boats at Goolwa have been refloated, together with the property prices at the Hindmarsh Island marina. Big real estate has got water; the farmers, fishermen and environmentalists have not. How very surprising.

Our national river, the 'mighty' Murray, still fails to reach the sea. At its end it's unlike any dictionary river: there is no copious stream, there is no water flowing, there is no channel reaching the sea. There are only short-term solutions, piling one onto another. The river has always been a metaphor. Now another summer is bearing down upon it, and I fear what it may come to represent to future Australians.

References

There are two ways to discover the Murray-Darling. The first is to grab a few maps, jump in the car and head off. The second is to stay at home, read some books and surf the Net. It's a lot less expensive and time consuming, yet you're able to glean an amazing amount of information. Think of it as a virtual journey. My real-world journey in *The River* was guided in part by these books and websites.

A good place to start is the Murray-Darling Basin Authority's website: www.mdba.gov.au. Here you can find out how the system is managed, plus up-to-date information on how much water is in Dartmouth, Hume and so on. And you can check out the latest drought update to learn how the system is coping.

If you want to know what's coming down the track, check out the Bureau of Meteorology site: www.bom.gov.au. There are three-month predictions on where the weather is heading, plus up-to-date analysis of the likelihood of drought-inducing El Niño weather patterns.

At www.csiro.gov.au there's comprehensive information on climate change and the strategies being developed to adapt to it. There's also information about the CSIRO's sustainable yields project, which attempts to determine how much water can be extracted from our river systems. The Wentworth Group of Concerned Scientists site at www.wentworthgroup.org provides insight into the current thinking of leading Australian researchers on the state of the climate, the rivers and the land.

REFERENCES

The Garnaut Report and other publications can be found at www.garnautreview.org.au. For a global perspective, try the website of the Intergovernmental Panel on Climate Change at www.ipcc.ch.

Each state and territory has its own multitiered water bureaucracy, with a plethora of websites. There are the water departments and usually a corporation responsible for water delivery, plus the irrigation companies and individual catchment authorities. The websites www.g-mwater.com.au, www.murrayirrigation.com.au and www.mirrigation.com.au provide insights into the world of irrigators. For those more interested in the environment, the Australian Conservation Foundation has mounted a healthy rivers campaign: see www.acfonline.org.au.

The tourist website www.murrayriver.com.au contains some entertaining insights, while www.backobourke.com.au fills something of the same role for the Darling River.

Useful books include Ticky Fullerton's *Watershed* (ABC Books, 2001), Paul Sinclair's *The Murray* (MUP, 2001), Asa Wahlquist's *Thirsty Country* (Jacana Books, 2008) and Tim Flannery's *The Weather Makers* (Text, 2006). The four-part television documentary *Two Men in a Tinny* is available on DVD from the ABC. And don't forget Henry Lawson. His complete works have been published as a two-volume set: *A Camp Fire Yarn* and *A Fantasy of Man* (Lansdowne, 1984).

Acknowledgements

I had always thought writing a rather solitary profession until I undertook this, my first book. Along the way, I realised how much I depended on the support and goodwill of friends, family and total strangers.

First and foremost I must thank my wonderful wife, Tomoko Akami, who always found time away from writing her own books and the demands of her own career to give love and encouragement. Our children, Cameron and Elena, have been wonderful; they never resented my absences, but found joy in the prospect of Dad writing a book.

Thanks also to my mum and dad, Glenys and Kevin, for giving the same love, encouragement and support they have always given, and for recounting their days living in the Victorian Mallee.

A special thanks to Louise Adler from MUP for seeing the potential of *The River*. There should be a special place reserved in heaven for publishers willing to take a punt on first-time authors. Thanks also to MUP's Executive Publisher, Foong Ling Kong, who oversaw the book with interest and understanding, and to Penny Mansley, Terri King and Maria O'Dwyer. Thanks also to copyeditor Paul Smitz, whose meticulous attention to detail converted a sloppy manuscript of typos and poor punctuation into something halfway readable.

I am grateful for the gracious advice and support of friends, particularly those who were kind enough to read my first draft and offer their advice: Paul Daley, Michael Brissenden, Katherine Murphy and Chas Savage.

From the very beginning, I was determined that this book be free of the usual suspects: the politicians and experts who populate the daily media. But that doesn't mean I didn't consult any of them. My thanks to Wendy Craik and, especially, Don Blackmore, both former heads of the Murray–Darling Basin Commission. Thanks also to Arlene Buchan from the Australian Conservation Foundation. Blame me for any errors or fanciful conclusions, not them.

Finally, this is a book that relies almost entirely on the cooperation and input of all those strangers I met during my travels, who freely donated their time, their insights and their memories. Many are mentioned by name in this book; others are not. My thanks to them all in equal measure.